T0325824

INTRODUCTION TO ACCELERATOR DYNAMICS

How does a particle accelerator work? The most direct and intuitive answer focuses on the dynamics of single particles as they travel through an accelerator. Particle accelerators are becoming ever more sophisticated and diverse, from the Large Hadron Collider (LHC) at CERN to multi-MW linear accelerators and small medical synchrotrons. This self-contained book presents a pedagogical account of the important field of accelerator physics, which has grown rapidly since its inception in the latter half of the last century. Key topics covered include the physics of particle acceleration, collision and beam dynamics and the engineering considerations intrinsic to the effective construction and operation of particle accelerators. By drawing direct connections between accelerator technology and the parallel development of computational capability, this book offers an accessible introduction to this exciting field at a level appropriate for advanced undergraduate and graduate students, accelerator scientists and engineers.

STEPHEN PEGGS is a senior scientist at Brookhaven National Laboratory (BNL) and an adjunct professor in physics at Stony Brook University, where he was closely involved in building and commissioning the Relativistic Heavy Ion Collider (RHIC). He has worked internationally on a variety of accelerators, including the Cornell Electron Storage Ring, the Super Proton Synchrotron (SPS) collider at CERN, the Superconducting Super Collider in Texas, the Tevatron and the Main Injector at Fermilab and the European Spallation Source in Sweden. He is a fellow of the American Physical Society.

TODD SATOGATA is a senior physicist at the Center for Advanced Studies of Accelerators at Jefferson Lab and a Jefferson Lab professor at the Center for Accelerator Science at Old Dominion University.

During his career, he has worked on the commissioning, design, building and operation of the Relativistic Heavy Ion Collider at BNL and commissioning and operation of the 12 GeV CEBAF upgrade at Jefferson Lab, in addition to developing medical accelerators, proton beam imaging techniques and accelerator control systems.

INTRODUCTION TO ACCELERATOR DYNAMICS

STEPHEN PEGGS
Brookhaven National Laboratory, New York

TODD SATOGATA
Jefferson Lab

Shaftesbury Road, Cambridge CB2 8EA, United Kingdom

One Liberty Plaza, 20th Floor, New York, NY 10006, USA

477 Williamstown Road, Port Melbourne, VIC 3207, Australia

314–321, 3rd Floor, Plot 3, Splendor Forum, Jasola District Centre, New Delhi – 110025, India

103 Penang Road, #05–06/07, Visioncrest Commercial, Singapore 238467

Cambridge University Press is part of Cambridge University Press & Assessment,
a department of the University of Cambridge.

We share the University's mission to contribute to society through the pursuit of
education, learning and research at the highest international levels of excellence.

www.cambridge.org
Information on this title: www.cambridge.org/9781107132849

DOI: 10.1017/9781316459300

First published 2017

A catalogue record for this publication is available from the British Library

ISBN 978-1-107-13284-9 Hardback

To
Rachel and Leah

Contents

Preface

This book addresses the single particle dynamics heart of the question; 'How does an accelerator work?', for readers who are accelerator users and operators, who are accelerator physicists or who are interested in real-world linear and nonlinear difference systems. The reader might be a synchrotron light source user, a collider experimentalist, a medical accelerator operator, an engineer in a beam instrumentation group or a controls professional in industry.

The level of the discussion is appropriate for graduate students, final-year undergraduates and practicing accelerator professionals. This is not an exhaustively complete reference handbook, at a high technical level. Rather, it is a pedagogical introduction to the subject, telling a self-contained and accurate story about a field of physics that was born and grew rapidly in the second half of the twentieth century, and which continues to mature by leaps and bounds in the twenty first. The treatment is rigorous enough to be accurate and useful, without letting unnecessary detail obscure the central theme. Deeper investigations of the 'back-stories' are left to other sources.

The central theme is that repetitive motion through an accelerator is a natural, convenient and well-motivated introduction to the generic linear and nonlinear behaviour of highly iterated difference systems. A circular accelerator – or even a linear accelerator – is one answer to another question: *'How do difference systems manifest themselves in the real world?'*

1
Introduction

1.1 Differences or Differentials?

Computers have been used to model physical systems from their earliest days. The behaviour of these models is subtly but profoundly affected by the sequential and discrete nature of all digital computers, in which one step in an algorithm follows another, just as a charged particle follows one turn after another around a circular accelerator. Numerical tools – computers – make dynamical problems look like difference equations. Analytic tools – most pre-computer mathematics – make problems look like differential equations. The best choice of representation depends on the nature of the particular system being modelled, and on the questions that are to be answered.

Figure 1.1 illustrates a simple nonlinear system – a gravity pendulum with unit length – that is inherently continuous in time, and which is accurately represented by the differential equation

$$\theta'' = -g \sin(\theta) \tag{1.1}$$

where a prime indicates differentiation with respect to time, and g is the acceleration due to gravity. Numerically, this motion is simulated by breaking the motion into two steps that are iterated repeatedly. In the first step, the pendulum 'drifts' with a constant angular velocity θ' for a time Δt. In the second step, gravity is represented by a 'kick' impulse that instantaneously changes the angular velocity. This is written in pseudo-code as

$$\begin{aligned} &\text{until finished } \{ \\ &\qquad \theta = \theta \;+\; \theta' \,.\, \Delta t \\ &\qquad \theta' = \theta' \;-\; g \sin(\theta) \,.\, \Delta t \\ &\} \end{aligned} \tag{1.2}$$

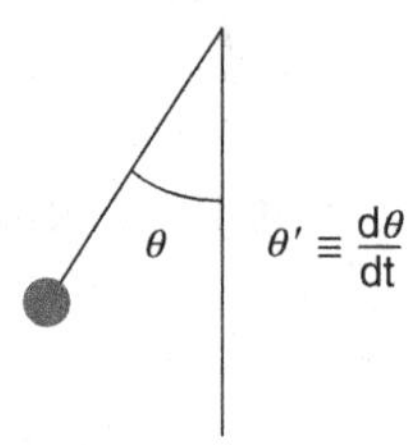

Figure 1.1 Phase space co-ordinates (θ, θ') for a unit length gravity pendulum.

Clearly, the differential and difference representations of a pendulum become identical as Δt approaches zero. Less clear is the potential for the numerical model to introduce artefacts into the simulated motion – behaviour that is not demonstrated by actual pendula.

Circular accelerators are inherently discrete in time, given the turn-by-turn nature of the progress of a test particle. It will be seen, in Chapter 4, that the *standard map* motion of Equation 1.2 (with g set equal to 1) also naturally describes the turn-by-turn longitudinal oscillations of a test particle, relative to an ideal particle at the centre of a bunch. Further, the standard map that naturally models many apparently independent nonlinear accelerator scenarios is a *standard* way to generate chaos, a phenomenon not demonstrated by gravity pendula.

Sometimes analytic solutions to differential equations are known. This is the case for the pendulum, a non-trivial nonlinear system (when large angles are considered) that was completely solved well before computers were invented. Analytic solutions to nonlinear *difference* systems are much rarer. Most traditional mathematical methods are implicitly continuous, and break down more or less completely when truly differential systems, like accelerators, are modelled.

Although computers and accelerators proliferated together, in the second half of the twentieth century, it took time to recognise the limitations of traditional continuous methods. For example, the difference map of Equation 1.2 can be forced into a differential representation by writing

$$\theta'' = -\sum_{n=1}^{\infty} \delta(t - n\Delta t)\ g\ \sin(\theta)\,.\,\Delta t \tag{1.3}$$

where the delta function $\delta()$ is non-zero every Δt units of time. The perceived advantage of this kind of representation of accelerators was that it enabled traditional Hamiltonian, Fourier and perturbation analysis tools to be used. It eventually became clear that traditional methods often break down more or less completely, and that numerical techniques are not just more convenient, but are also more fundamentally correct. Often, particles must be numerically tracked around a circular accelerator, turn-by-turn.

It is often fast and easy to model a physical system by writing a short program containing an approximate numerical model, even if the system is inherently continuous. Two deceptively simple real-world systems – a tape drive and a betatron – are discussed in Exercises 2.8 and 7.5.

1.2 Phase Space Co-ordinates

Time advances discretely in a circular accelerator, labelled by the integer turn number n.

Figure 1.2 illustrates the observation of the horizontal displacement x and the tangent $x' = dx/ds$ of a test particle as it passes a reference point at $s = s_0$ on the circumference of an accelerator. Here $0 \leq s < C$ measures the location of objects – magnets and beam diagnostics, as well as test particles – projected onto the 'design orbit' that is followed by an idealised test particle as it circulates. Usually, but not always, the design orbit goes down the centre of a perfectly aligned beampipe. In general the angle of the test particle is small

$$x' = \tan(\psi) \approx \psi \sim 10^{-4} \tag{1.4}$$

and there is little practical distinction between x' and the horizontal angle ψ with respect to the beampipe centre line.

Figure 1.3 illustrates the first four observations of the horizontal phase space motion at the reference point $s = s_0$, in the infinite series $(x, x')_n$; $n = 0, 1, \ldots, \infty$.

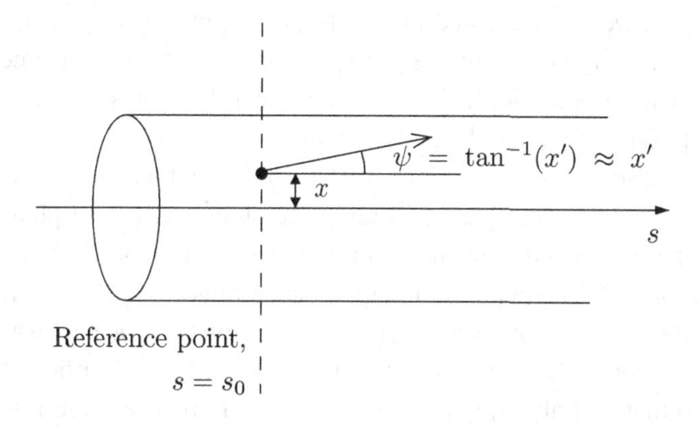

Figure 1.2 Horizontal phase space co-ordinates (x, x') for a test particle moving through a beampipe, and repeatedly passing a reference point at $s = s_0$, turn-by-turn.

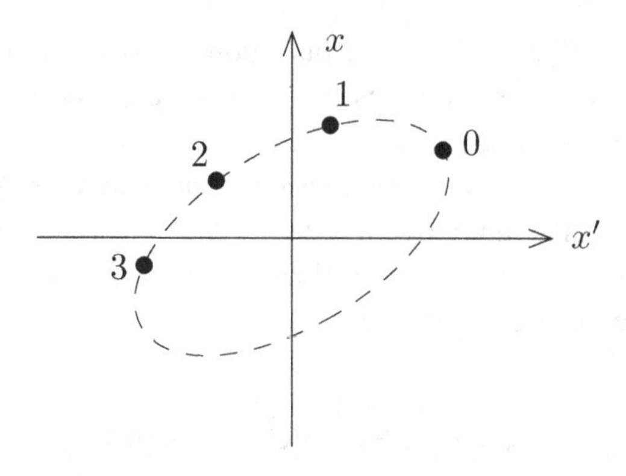

Figure 1.3 Motion of a test particle in horizontal phase space $(x, x')_n$ as it advances for 3 turns (with $n = 0, 1, 2$ and 3) around an accelerator. This Poincaré surface of section is observed at a reference point at $s = s_0$.

In this idealised case the co-ordinate pairs $(x, x')_n$ all lie on an ellipse in phase space, corresponding (as will be seen) to an ideal situation that is completely linear and stable. In practice, the phase space motion observed in this *Poincaré surface of section* is at least a little nonlinear, and the ellipse is at least a little distorted. If conditions are bad enough, the test particle collides with the beampipe wall or some other obstacle when $|x| > x_{max}$, at a time n that falls far short of infinity.

This raises a very practical question: under what circumstances do test particles remain within the beampipe for a sufficiently long time? A particle moving close to the speed of light c in an accelerator with a circumference C of 1 kilometre has a revolution frequency $f_{rev} \approx c/C$ of about 0.3 MHz. If this beam is to be stored for about a day, then we are interested in stability over times as long as $n \approx 3 \times 10^{10}$ turns, a timescale comparable to the age of the solar system, measured in periods (years) of the earth's rotation. In fact, the nonlinear dynamics of solar system and test particle stability are quite closely related.

Phase space descriptions of motion are also convenient when time is continuous. Figure 1.4 illustrates the motion of a gravity pendulum in (θ, θ') phase space, for the unit length gravity pendulum shown in Figure 1.1, and modelled by the standard map of Equation 1.2. The pendulum angle is constrained to lie in the range $-\pi \leq \theta < \pi$ by subtracting or adding 2π if it winds forwards (or backwards) beyond $\theta = \pm\pi$. This graphic was generated by reducing the time step sufficiently far in the limit $\Delta t \to 0$ that neighbouring dots appear to overlap. The more interesting and relevant case when Δt has finite values that correspond to accelerator experience is discussed in Chapter 4.

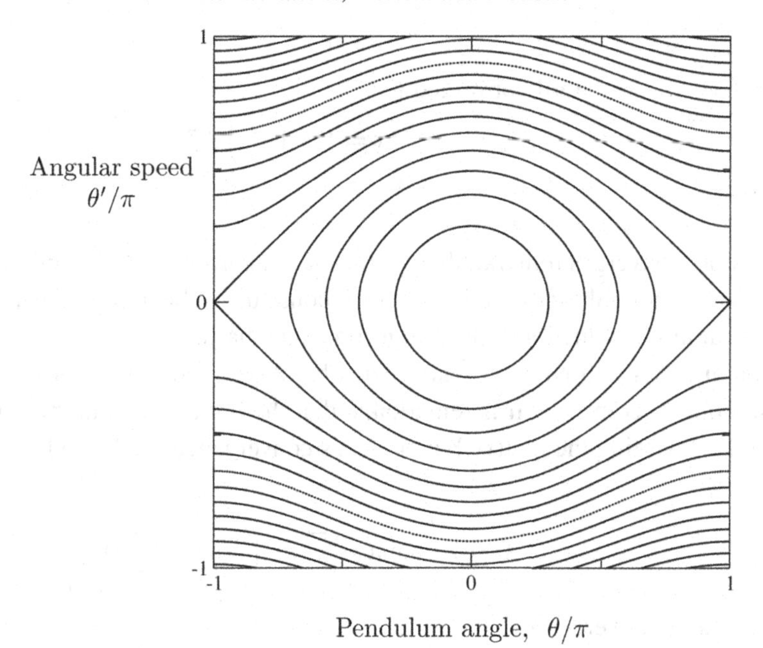

Figure 1.4 Motion of a unit length gravity pendulum (with $g = 1$) in (θ, θ') phase space. The pendulum angle θ is reduced to always lie in the range from $-\pi$ to $+\pi$, while the angular speed θ' is unconstrained: the pendulum can wind forwards or backwards.

1.3 Iterations, Ancient and Modern

Although computers are relatively recent, iterative algorithms have been around for a long time. One well-known example is the Newton search method for solving the equation $f(x) = 0$

$$x = \text{guess} \tag{1.5}$$

$$\text{until converged } \{$$

$$x = x - \frac{f(x)}{f'(x)}$$

$$\}$$

where $f'(x) = df/dx$ is the differential of the general function f. Algorithms that are specific examples of the general method precede Newton himself. For example, for centuries before the arrival of tables of logarithms and calculators, the square root of y was found as follows

$$
\begin{aligned}
& x = 1 \\
\text{until converged } \{ & \\
& x = \left(x + \frac{y}{x}\right) / 2 \\
\} &
\end{aligned}
\tag{1.6}
$$

This algorithm converges remarkably rapidly – the square root of 10 is correct to 10 decimal places after only six iterations – and it continues to be used in computerised mathematical library functions. Good algorithms die hard.

The initial guess in a Newton search must be close enough to the right answer for the solution to converge. It is remarkable that the initial guess in the following algorithm for inverting the matrix Y guarantees convergence [2, 37, 51].

$$
\begin{aligned}
& X = \tilde{Y} / \left(\max_i \sum_j |A_{ij}| \, . \, \max_j \sum_i |A_{ij}| \right) \\
\text{until converged } \{ & \\
& X = X + X(I - XY) \\
\} &
\end{aligned}
\tag{1.7}
$$

Here $\tilde{Y}$ is the transpose of Y, and I is the identity matrix. The most difficult requirement of this algorithm is that the user be able to multiply matrices. A program using this algorithm to invert matrices is not the fastest, but it is surely one of the easiest to understand and construct. These are two essential elements of good software engineering.

In solving a problem numerically there is often a choice between a more or less direct transcription of an analytic method (for example, for matrix inversion), and a less traditional iterative method. An iterative approach is often easier to understand, and is more concise and flexible. This is true for solar system and accelerator dynamics, as well as for computational algorithms.

1.4 Accelerator History: The Two Golden Ages

The first golden age of accelerator construction and innovation began after the second world war, almost exclusively in support of particle physics experimentation. In those days an accelerator – typically a synchrotron – accelerated electrons, protons, or charged ions to relativistic speeds before diverting the beam onto a fixed target that may have been composed mainly of protons (in a liquid hydrogen target), or perhaps of heavier nuclei. Accelerator technology advanced rapidly in pursuit of ever higher centre-of-mass collisions, in order to create new, heavier, particles.

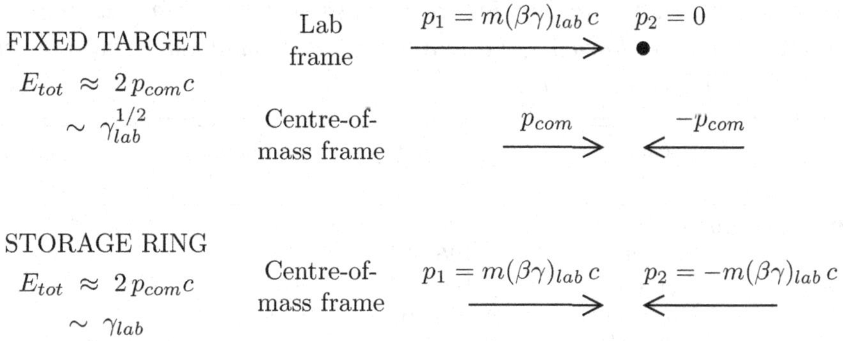

FIXED TARGET

$E_{tot} \approx 2\,p_{com}c$

$\sim \gamma_{lab}^{1/2}$

Lab frame: $p_1 = m(\beta\gamma)_{lab}\,c$ ⟶ $p_2 = 0$ •

Centre-of-mass frame: p_{com} ⟶ ⟵ $-p_{com}$

STORAGE RING

$E_{tot} \approx 2\,p_{com}c$

$\sim \gamma_{lab}$

Centre-of-mass frame: $p_1 = m(\beta\gamma)_{lab}\,c$ ⟶ ⟵ $p_2 = -m(\beta\gamma)_{lab}\,c$

Figure 1.5 Storage rings have a significant advantage over fixed target accelerators, because the total useful centre-of-mass energy E_{tot} scales like γ with the beam energy, in comparison to only $\gamma^{1/2}$ in fixed target accelerators, for large values of γ. The lab frame and the centre-of-mass frame are the same for a (symmetric) storage ring collider.

Storage ring colliders deliver centre-of-mass energies that scale like the relativistic factor γ of the beam, rather than the $\gamma^{1/2}$ scaling inherent to fixed target accelerators. This is illustrated in Figure 1.5. The enhanced efficiency of storage rings comes at the cost of the added complexity of ensuring that two counter-rotating beams pass through each other at (only) a few collision points. Electron storage rings – with intrinsically larger values of γ – began operation in the 1960s, followed by proton, ion and *ep* colliders in the 1980s and beyond, as summarised by Table 1.1.

Electron–positron collisions are experimentally more desirable than electron–electron collisions, because of the cancellation of quantum numbers in the collision. Practical e^+e^- colliders began to operate in the 1970s, while copious supplies of anti-protons only became available (thanks to the invention of stochastic cooling) for proton–antiproton $p\bar{p}$ colliders in the 1980s. Particle–antiparticle collisions are fortunately possible in a single storage ring, thanks to an important feature of the Lorentz equation

$$\vec{F} = q\,(\vec{E} + \vec{v} \times \vec{B}) \tag{1.8}$$

where q is the charge of a particle with a velocity $\vec{v}$ passing through an electric field $\vec{E}$ and magnetic field $\vec{B}$. If there are no electric fields ($\vec{E} = 0$), then the same force $\vec{F}$ is experienced at the same location by counter-rotating particles and anti-particles, under the transformation

$$\begin{aligned} q &\to -q \\ \vec{v} &\to -\vec{v} \end{aligned} \tag{1.9}$$

Table 1.1 *A partial list of storage ring colliders, showing the historical progression from a proliferation of low-energy electron rings to a small number of very high-energy hadron rings (Wikipedia).*

Name	Location	Operating years	Energy [GeV]	Comment
Electron–positron rings				
AdA	Frascati	1961-64	0.25	e^-e^-
P-S	Stanford	1962-67	0.30	e^-e^-
VEP-1	Novosibirsk	1964-68	0.13	e^-e^-
SPEAR	Stanford	1972-90	3	
DORIS	Hamburg	1974-93	5	
CESR, -c	Ithaca	1979-08	6	
LEP	Geneva	1989-00	104	
BEPC	Beijing	1989-04	2.2	
PEP-II	Stanford	1998-08	9, 3.5	Asymmetric
KEKB	Tsukuba	1999-09	8, 3.5	Asymmetric
Hadron rings				
ISR	Geneva	1971-84	32	pp
Sp$\bar{p}$S	Geneva	1981-84	270	$p\bar{p}$
HERA	Hamburg	1992-07	28, 920	ep
Tevatron	Batavia	1992-11	980	$p\bar{p}$
RHIC	Upton	2000-	250, 100	$\vec{p}$, ions
LHC	Geneva	2008-	4000	pp, ions

Thus, particle and anti-particle follow the same trajectory in opposite directions, and bunches of counter-rotating particles are guaranteed to collide (pass through each other) head-on. Further, perturbative electrostatic fields can be used to manipulate a beam of many bunches to collide at desirable locations in the middle of particle detectors, but not at other locations around the arcs of the storage ring.

Luminosity L is the second key measure of storage ring performance, in addition to centre-of-mass energy. It measures the ability of particle collisions to generate events with a cross section σ, through the equation

$$R = L\sigma \tag{1.10}$$

where R is the average rate of event generation. In the simple case of constant cross section Gaussian beams, the luminosity is

$$L = f_{rev}\, M\, \frac{N^2}{4\pi\,\sigma_H^*\sigma_V^*} \tag{1.11}$$

where f_{rev} is the revolution frequency of M identical bunches, each with N particles per bunch and with horizontal and vertical RMS transverse sizes of σ_H^* and σ_V^* at the collision point. A third quantity of interest is the total current in each beam,

$$I = Ze f_{rev} NM \tag{1.12}$$

where Z is the particle charge (1 for protons, electrons and positrons, and 79 for gold ions). Finally, the energy stored in each beam is

$$U = NM A\, m_0 c^2 \gamma \tag{1.13}$$

where for ions $m_0 = 1.661 \times 10^{-27}$ kg is the atomic mass unit, and A is the atomic weight (1.008 for hydrogen and 196.967 for gold). See Exercise 1.2 for a discussion of typical parameters of the Relativistic Heavy Ion Collider, RHIC [18].

The trend to ever-higher luminosities is illustrated for gold ion beam collisions in the RHIC two-ring collider in Figure 1.6, for the period 2007 to 2014. Equation 1.11 shows that there are three main paths to higher luminosity: more bunches M, more particles N and smaller beam sizes σ^*. The number of bunches is increased by separating the beams into two rings, so that counter-circulating bunches do not see and feel the electromagnetic interactions between each other that are inevitable, even if the beams are transversely separated, in a single ring.

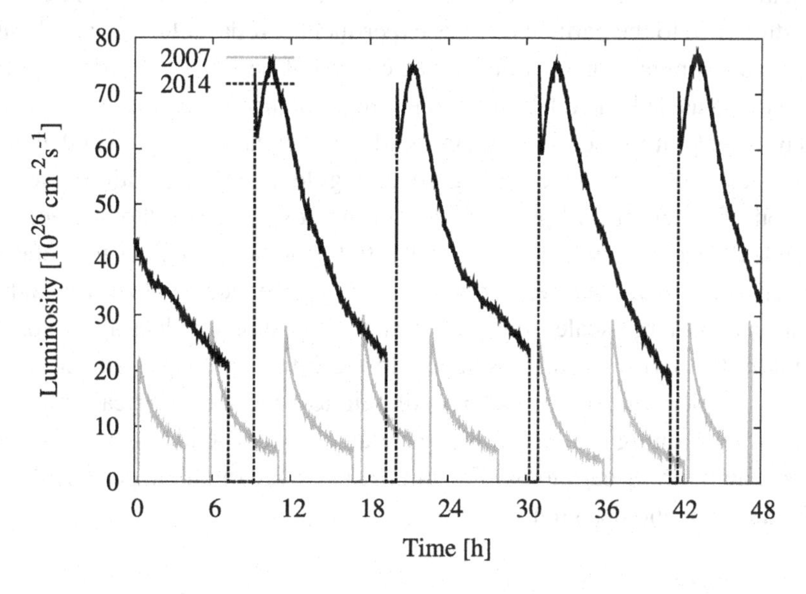

Figure 1.6 Luminosity versus time in the Relativistic Heavy Ion Collider, and its enhancement from 2007 to 2014. The grey curve shows two days of good running with gold-on-gold collisions in 2007, in which shorter stores show a classic exponential luminosity decay. The black curve shows longer stores in 2014, with approximately 90% of the gold ions productively 'burnt-off' in luminosity production. The luminosity grows initially, thanks to increasing bunch density due to transverse and longitudinal cooling. (Courtesy of W. Fischer [15].)

Increasing N and decreasing σ^* both increase the transverse density of a bunch, enhancing the transverse electromagnetic fields that accompany it, and exacerbating the violent impulse visited on counter-rotating test particles in a second bunch that passes through the first bunch. The nonlinear dynamics of this beam–beam effect (discussed in Chapter 15) limit the maximum achievable beam density, and therefore also the luminosity. Before discussing such nonlinear dynamics, however, it is first necessary to establish the linear language of beam optics that (for example) describes and enables the controlled manipulation of beam sizes like σ^*.

The second golden age of accelerator development began with the practical application of the fact that an electron that is forced to bend its trajectory radiates photons, typically X-rays, while moving through an accelerator. The strength of this synchrotron radiation increases inversely with a high power of the particle mass m, which is almost 2,000 times smaller for an electron than for a proton. For most practical purposes, GeV-range electrons radiate copiously, while protons and ions do not radiate 'at all'.

Synchrotron radiation was a curiosity in the early days of electron particle physics experimentation – even a nuisance, because the power emitted from the beam must be replaced, and paid for. Later, synchrotron radiation was used parasitically, to perform surface science, biological and other experiments, with little disturbance to the particle physics experiments. Today, electron accelerators – both circular storage rings and linear free electron lasers – are designed, built and commissioned as dedicated facilities solely for synchrotron light uses.

Today, accelerators continue their rapid diversification far beyond their particle physics roots, in the on-going second golden age. In addition to a new generation of synchrotron light sources, cutting edge and overlapping accelerator research and development is active in pursuit of room-size medical accelerators for proton and ion beam cancer therapy, medium-size accelerators for industrial applications and large-scale high-power MW-class superconducting proton linear accelerators (linacs) for neutron science. In the future, perhaps multi-MW proton sources in accelerator driven reactors will help to generate electrical power, at the same time as burning nuclear waste generated by conventional uranium power reactors. The future is hard to predict, but the language of single particle dynamics helps to describe and design it.

Exercises

1.1 The *logistic map* advances a population x from generation n to generation $n+1$ through the equation

$$x_{n+1} = \alpha\, x_n(1 - x_n) \tag{1.14}$$

where the control parameter α is in the range from 0 to 4, and x is in the range from 0 to 1.

a) How does x evolve as a function of time?
b) Can the description of this evolution be generalised for a range of initial values of x?
c) Under what circumstances – for example, for what ranges of α – can the logistic map be solved analytically?

1.2 Suppose that the peak luminosities shown in Figure 1.6 are generated when RHIC stores 120 identical bunches of 10^9 gold ions at an energy of 100 GeV/u in each of its two rings, distributed around the circumference of 3.834 km.

a) What is the approximate transverse size of the beam, assuming it to be round?
b) If RHIC instead stores polarized proton bunches with the same charge per bunch at an energy of 250 GeV, and with the same transverse size, what is the luminosity?
c) What is the beam current in each ring for the two cases – gold ions and protons?
d) What is the stored energy in each beam?

1.3 In one dimension the work done by a force F acting through a distance ds is $dE = F\,ds$. Show *directly* that the energy gain of a particle of mass m that increases its Lorentz factor by $\Delta\gamma$ is

$$\Delta E = \Delta\gamma\, mc^2 \tag{1.15}$$

where the rest energy of the particle is $E_0 = mc^2$. From this it follows that $E = E_0\gamma$. Use this result to show that

$$E^2 = p^2c^2 + m^2c^4 \tag{1.16}$$

1.4 Two similar particles collide head-on, each with a centre-of-mass energy of $mc^2\gamma_{cm}$. Show that this is equivalent to a collision in which one particle is at rest and the other has an energy of $mc^2\gamma$, where

$$\gamma = 2\gamma_{cm}^2 - 1 \tag{1.17}$$

1.5 The Superconducting Super Collider (SSC) was to have had head-on collisions between 20 TeV proton bunches in two rings. What energy would be required for fixed target accelerator protons striking a liquid hydrogen target, to produce the same centre-of-mass collision energy?

1.6 The electron–proton collider HERA operated with two rings: one storing 30 GeV electrons, and the other storing 800 GeV protons. What was the center-of-mass collision energy?

1.7 Show that the incremental increase in energy dE is related to the incremental increase in momentum dp through

$$\frac{dE}{E} = \beta^2 \frac{dp}{p} \tag{1.18}$$

1.8 Suppose that two beams of particles with Gaussian distributions collide head-on, except for a horizontal offset b. Show that the luminosity is reduced by a factor of $\exp(-b^2/4\sigma_x^{*2})$, where σ_x^* is the horizontal root mean square size of both beams. What practical use can be made of this dependence?

1.9 The transverse sizes of two beams in a collider are constant in time, but the two currents decay with an exponential time constant of τ. If it takes time T from dumping the remaining stored beams to bringing freshly injected full intensity beams back into collision, what is the optimum time T_{store} to take data, before dumping beams and repeating the operational cycle?

1.10 Consider a charged pion decaying into a muon plus an antineutrino:

$$\pi^- \rightarrow \mu^- + \bar{\nu}_\mu$$

Use $m_{\pi^\pm} = 140$ MeV/c^2, $m_\mu = 106$ MeV/c^2, and $m_{\bar{\nu}} \approx 0$.

a) In the pion rest frame, what are the energies and momenta of the muon and anti-neutrino?

b) Neutrinos have a tiny mass – assume $m_{\bar{\nu}}c^2 = 0.01$ eV. How high must the pion beam energy be, in order to produce some neutrinos at rest?

c) A pion with total energy $m_\pi c^2 \gamma$ decays. Find an expression that relates the muon angles θ_μ and θ_μ^*, defined in Figure 1.7.

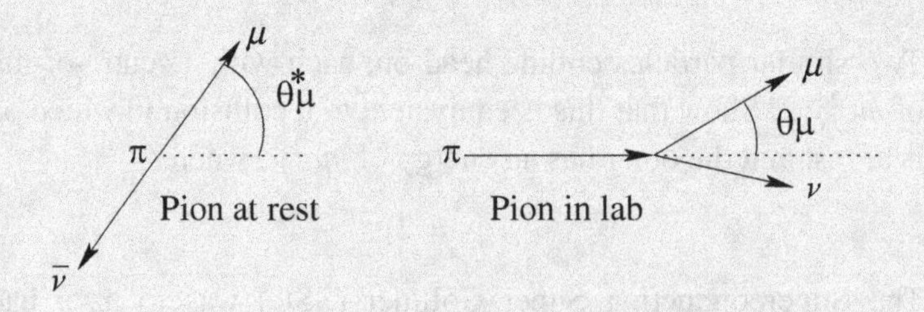

Figure 1.7 A pion decays into a muon and an antineutrino in the pion rest frame (left) and in the lab frame (right).

2
Linear Motion

The magnets in a circular accelerator are placed in a sequence, or lattice, that is intended to return an ideal test particle, one turn later, to the same location from where it was launched. Usually, this ideal location – the design orbit – is at the centre of the vacuum beampipe, as illustrated in Figure 2.1. It can be shown mathematically that, with static fields ($dB/dt = dE/dt = 0$), there is one orbit – the closed orbit – that exactly repeats itself, turn after turn. In practice, the design orbit and the closed orbit are not the same, because of the inevitable presence of errors of various sorts, such as magnet misalignments.

Errors, and their correction, are discussed in Chapters 6 and 8. For now, it is assumed that the closed orbit and the design orbit are identical – there are no errors. Even so, it is necessary to consider non-ideal test particles that deviate at least a little from the design orbit. Particles circulate the accelerator in bunches, perhaps 10^9 per bunch. Their stability must be guaranteed as they oscillate horizontally and vertically, relative to an ideal particle at the centre of a bunch. Figure 2.2 shows the right-handed co-ordinate system that is commonly used to describe such betatron oscillations, with x and y denoting the horizontal and vertical displacements of the test particles, respectively.

2.1 Stable Oscillations

The horizontal displacement of a stable particle at a reference point on turn n is written

$$x(n) = a \sin(2\pi\, Q_x n + \phi_0) \qquad (2.1)$$

where a and ϕ_0 are the amplitude and the initial phase of the betatron oscillation. The fractional part of the horizontal betatron tune Q_x is typically not small in practice – it is 0.155 in the example shown in Figure 2.3 – and the integer part of Q_x is usually much greater than 1. An arbitrary integer multiple of 2π can be added

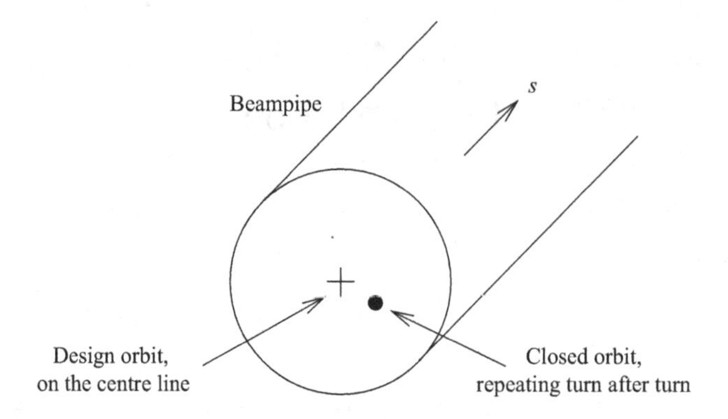

Figure 2.1 Design and closed orbits at a reference point in a synchrotron. The closed orbit is displaced from the design orbit by inevitable errors of various sorts – in magnet construction, alignment and operation. These errors are controlled in manufacture and assembly, and corrected in real time, to bring the closed orbit sufficiently close to the beampipe centre.

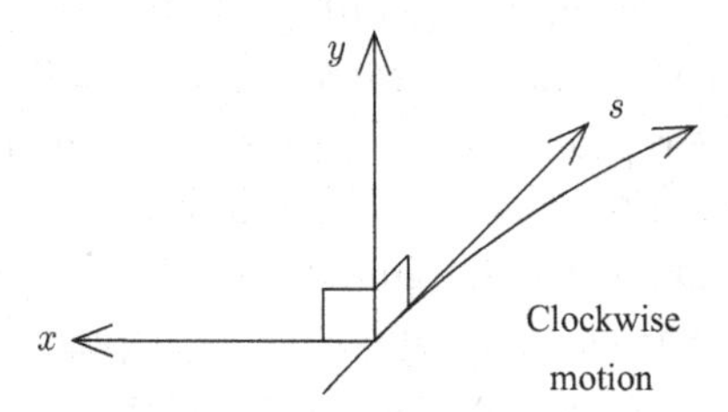

Figure 2.2 The right-handed co-ordinate system (x, y, s) that is commonly used for clockwise motion around an accelerator.

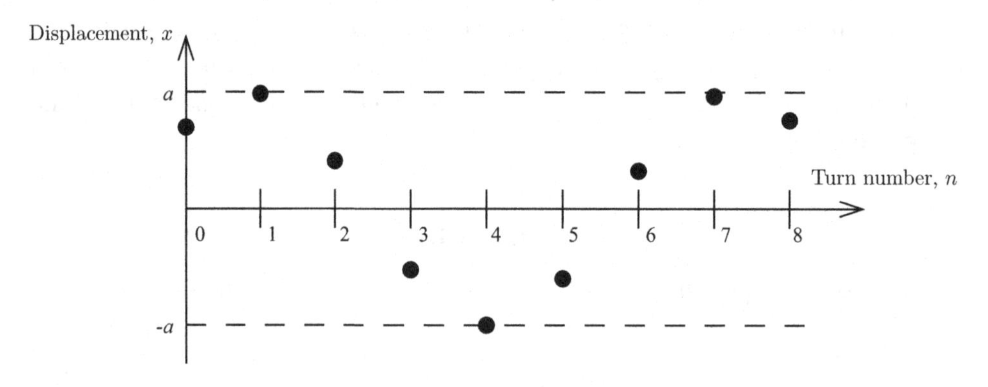

Figure 2.3 Horizontal betatron oscillations about the closed orbit, with an amplitude a and a fractional betatron tune of $\mathrm{mod}\ (Q_x, 1) = 0.155$.

to the argument of the sine in Equation 2.1, and so the integer component does not matter when only a single reference point is being considered. This accounting must be done correctly, however, when multiple reference points are used – for example, when considering the linear motion from one reference point to another around a storage ring.

Longitudinal stability also matters. The small synchrotron oscillation of a test particle towards the front and back of a bunch is written

$$z(n) = a_z \sin(2\pi\, Q_s\, n) \tag{2.2}$$

where z is the longitudinal displacement relative to the bunch centre. In contrast to horizontal (and vertical) betatron motion, the synchrotron tune is usually very small

$$0 < Q_s \ll 1 \tag{2.3}$$

except in extreme cases with very strong synchrotron radiation.

The longitudinal restoring force that provides stability comes from a small number of radio frequency (RF) cavities – perhaps only one – that together occupy an almost negligible fraction of the accelerator circumference. The representation of RF cavities by a single thin element – a delta function – is illustrated in Figure 2.4, which also indicates that bunches are usually needle-like: much longer than they are wide. Bunch lengths vary significantly, especially between electron and proton accelerators. Transverse sizes vary significantly between different azimuthal locations in a particular accelerator, while the bunch length is essentially constant.

All six phase space dimensions matter, not just the three displacement dimensions x, y and z. For example, a test particle on the closed orbit of an ideal accelerator has

$$(x, x', y, y', z, \delta) = 0 \tag{2.4}$$

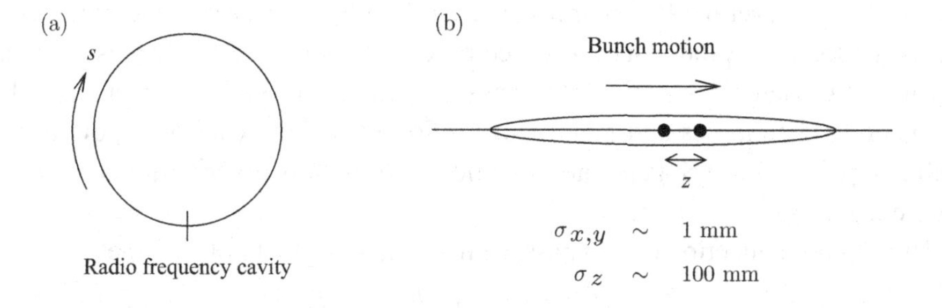

Figure 2.4 Longitudinal stability and typical beam sizes. The RF cavities that apply a longitudinal restoring force are often well approximated by a single thin element, as in (a). The needle-like bunch is usually a lot longer than it is wide. For example, the Relativistic Heavy Ion Collider (RHIC) proton bunch sketched in (b) has root mean square sizes of $(\sigma_x, \sigma_y, \sigma_z) \approx (1, 1, 100)$ mm.

where x' and y' are horizontal and vertical angles, respectively, and $\delta = \Delta p/p$ is the relative momentum offset. Discussion of the intertwined relationships between longitudinal oscillations, RF cavities, acceleration, nonlinear magnets and synchrotron radiation is deferred, necessarily, in favour of an immediate derivation of transverse linear motion through magnets – optics – from first principles.

2.2 Transverse Motion through Magnets

We are interested in the behaviour of a test particle with a total charge q and an initial total momentum

$$p_0 = mc \cdot \beta\gamma = mv \cdot \gamma \tag{2.5}$$

that passes through a magnet, somewhere near the centre line, and is deflected by the Lorentz force

$$\vec{F} \equiv \frac{d\vec{p}}{dt} = q\,\vec{v} \times \vec{B} \tag{2.6}$$

Assume that the magnet aperture is a lot smaller than its length L, and that the magnet has a constant cross-section, so that the magnetic field is purely transverse and does not depend on s

$$\begin{aligned} B_x &= B_x(x, y) \\ B_y &= B_y(x, y) \\ B_s &= 0 \end{aligned} \tag{2.7}$$

Maxwell's equations

$$\nabla \cdot \vec{B} = 0 \tag{2.8}$$

$$\nabla \times \vec{B} = \mu_0 J \tag{2.9}$$

are satisfied by several different transverse field configurations in the region of interest within the beampipe, where the electric current density J is zero, as discussed in Chapter 6. Here, it is sufficient to consider just two types of magnets, and their magnetic fields: dipoles and quadrupoles. For example, a cartoon dipole magnet with two poles, N and S, generates a vertical field with B_y = constant, as shown in Figure 2.5a.

The angular deflection due to passage through a single dipole magnet is

$$\theta = \frac{\Delta p_\perp}{p_0} = \frac{1}{p_0}\frac{dp}{dt}\frac{L}{v} = \frac{L}{\rho} \tag{2.10}$$

where ρ is the bending radius of the trajectory. Many such magnets bend the design orbit through 2π radians in the horizontal plane of a circular accelerator, when arranged in a lattice that also includes quadrupoles.

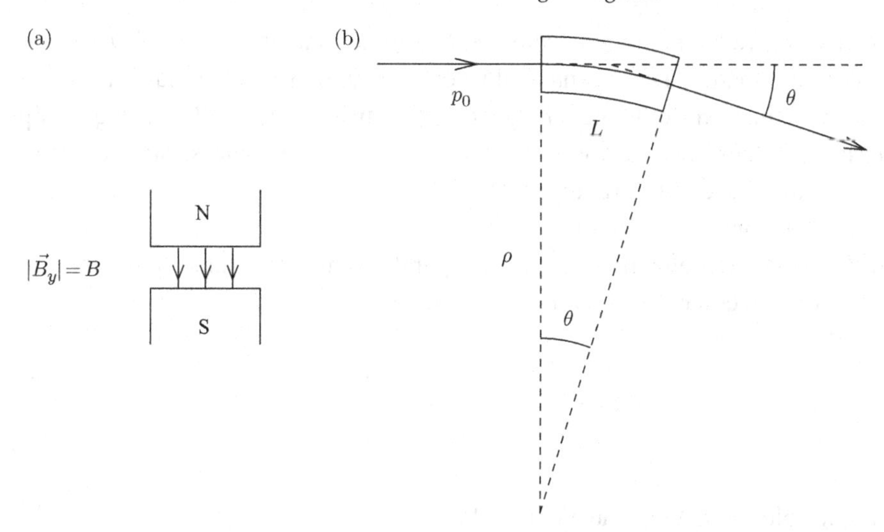

Figure 2.5 Geometry of motion through a dipole with a constant vertical field B. An end view is shown in (a). The top view in (b) shows a charged particle with initial momentum p_0 deflected by an angle θ during passage through a dipole magnet of length L.

Magnetic rigidity, a single quantity conventionally written as $(B\rho)$, encapsulates the fact that doubling the dipole field halves the bending radius. Equations 2.6 and 2.10, taken together, show that

$$G \equiv \frac{1}{\rho} = \frac{B}{(B\rho)} \tag{2.11}$$

where

$$(B\rho) = \frac{p_0}{|q|} \tag{2.12}$$

In general, magnetic rigidity converts between magnetic strength and geometric strength, no matter what the magnet style. A useful expression for a singly charged particle with $q = \pm e$

$$(B\rho)\ [\mathrm{Tm}] = 3.3356\, p_0\ [\mathrm{GeV/c}] \tag{2.13}$$

shows, for example, that a dipole with a field of 0.334 T bends a 1 GeV/c proton or electron in an arc of radius 10 m.

The field in a conventional iron-dominated magnet is controlled by the shape of the poles, so long as the maximum field at the poles is comfortably less than the saturation field of the iron, usually a little higher than 1 T. Higher fields – up to about 10 T, or perhaps somewhat more – are achieved by using superconducting wires or cables in a conductor-dominated magnet. The iron in a superconducting magnet is moved far enough from the beamline, and the beam, that saturation effects do not matter. The mechanical forces in superconducting magnets are much higher than in iron-dominated magnets, and must be controlled using complex material, mechanical and cryogenic technologies. Nonetheless, superconducting magnets have a second advantage of saving energy – electrical power is no longer dissipated in copper conductors.

If the magnetic fields in a magnet are purely transverse and do not vary with s, then Maxwell's equations 2.8 and 2.9 reduce to

$$\frac{\partial B_x}{\partial x} + \frac{\partial B_y}{\partial y} = 0 \tag{2.14}$$

$$\frac{\partial B_x}{\partial y} - \frac{\partial B_y}{\partial x} = 0 \tag{2.15}$$

A simple solution is the quadrupole field

$$B_x(y) = B'\,y \tag{2.16}$$
$$B_y(x) = B'\,x$$

where the horizontal and vertical fields vary linearly with the vertical and horizontal displacements, respectively, increasing at a constant rate B', the field gradient. This is illustrated for an iron-dominated magnet with four poles in Figure 2.6, which also shows how a particle entering a quadrupole magnet with a horizontal displacement x is deflected by a horizontal angle θ and is focused towards the axis.

If a quadrupole is relatively thin so that x is almost constant within the magnet, then to a good approximation

$$\theta = \frac{qB_y}{p_0} L = \frac{B'}{(B\rho)} Lx \tag{2.17}$$

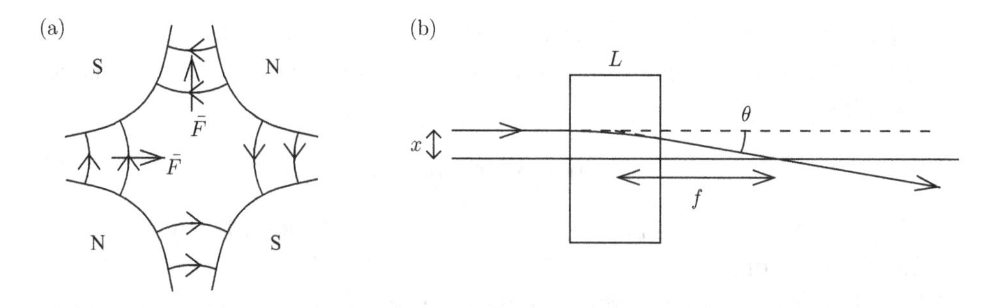

Figure 2.6 Geometry of motion through a iron quadrupole magnet, with forces $\vec{F}$ that focus in the horizontal plane, but which defocus in the vertical plane. An end view is shown in (a). The top view in (b) shows a particle with initial displacement x that crosses the axis at a focal distance f, after passage through a quadrupole of length L.

and the total angular deflection is proportional to the displacement

$$\theta = \frac{x}{f} \tag{2.18}$$

The integrated geometric strength of a thin quadruple – the inverse of the focal length f – is proportional to L

$$\frac{1}{f} = KL \tag{2.19}$$

where the local geometric strength of the quadrupole, K, is simply related to the field gradient B' by the magnetic rigidity

$$K = \text{sgn}(q)\,\frac{B'}{(B\rho)} \tag{2.20}$$

There is potential for confusion over signs, here, in part because rigidity is positive-definite while B', K and f may be negative. Horizontal motion through the thin quadrupole drawn in Figure 2.6, with positive K, is summarised by writing the phase space co-ordinates $(x, x')_2$ at the exit in terms of the co-ordinates $(x, x')_1$ at the entrance

$$\begin{aligned} x_2 &= x_1 \\ x_2' &= x_1' - \frac{x_1}{f} \end{aligned} \tag{2.21}$$

where f is positive.

2.3 Matrix Equations of Motion

Linear motion, magnet by magnet, is compactly represented by matrices. For example, the horizontal motion from s_1 to s_2 through a field-free drift of length L that is shown in Figure 2.7 becomes

$$\begin{aligned} x_2 &= x_1 + Lx_1' \\ x_2' &= x_1' \end{aligned} \tag{2.22}$$

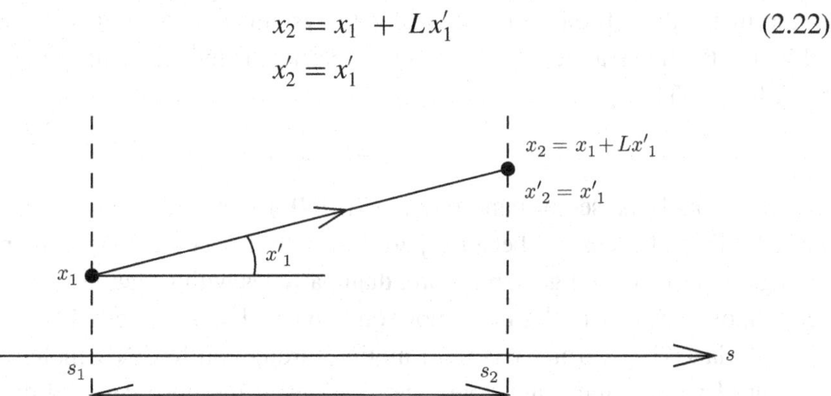

Figure 2.7 Transverse motion through a field-free drift of length L.

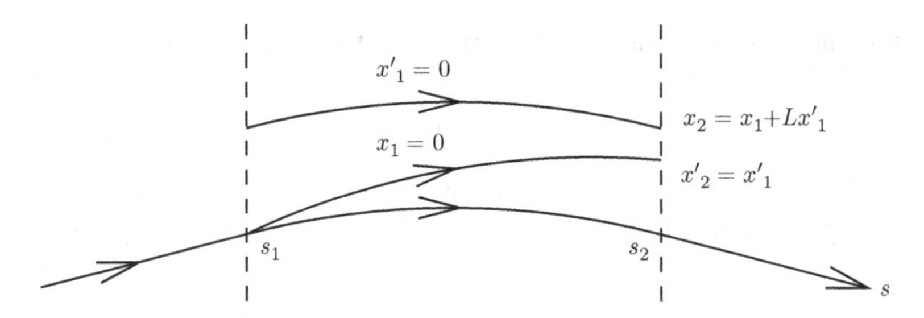

Figure 2.8 Transverse motion through a rectangular dipole. All trajectories have the same path length (to first order in the entrance angle x'_1), and so all have the same bend angle. The co-ordinate frame also rotates, and so the motion relative to the design orbit is the same as in a drift.

with an identical pair of expressions for vertical motion. The evolution of the 4D transverse phase space vector is therefore represented by M_{DRIFT}

$$\begin{pmatrix} x \\ x' \\ y \\ y' \end{pmatrix}_2 = M_{DRIFT} \begin{pmatrix} x \\ x' \\ y \\ y' \end{pmatrix}_1 \tag{2.23}$$

where

$$M_{DRIFT} = \begin{pmatrix} 1 & L & 0 & 0 \\ 0 & 1 & 0 & 0 \\ 0 & 0 & 1 & L \\ 0 & 0 & 0 & 1 \end{pmatrix} \tag{2.24}$$

Horizontal motion through a rectangular dipole (RBEND) is shown in Figure 2.8. The upstream and downstream magnet ends are parallel, and so the motion is same as through a drift, because the co-ordinate system rotates along with the design orbit, and the bend angle experienced by a particle is independent of the incoming co-ordinates. Thus,

$$M_{RBEND} = M_{DRIFT} \tag{2.25}$$

Motion through the sector bend dipole (SBEND) sketched in Figure 2.5 includes some horizontal focusing, because particles with a positive displacement x_1 pass through more field, and get bent more, than particles with negative x_1. Thus, sector bend dipoles focus particles in the horizontal plane, but act exactly like drifts in the vertical plane. The corollary is true: motion through an RBEND includes a small amount of vertical focusing. Pole-face rotations at the upstream and downstream ends of a dipole must always be considered carefully, even though the strength of the effect is small for dipoles in a large accelerator, with small bend angles.

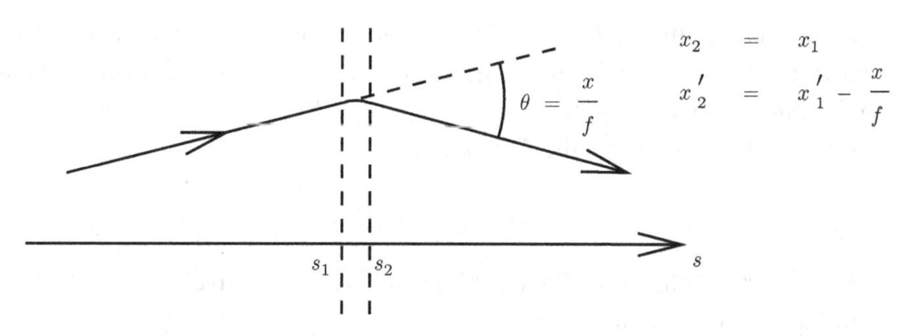

Figure 2.9 Transverse motion through a thin quadrupole that is much shorter than its focal length.

Transverse motion through a thin quadrupole with $L \ll |f|$, shown in Figure 2.9, is represented by the matrix

$$M_{THIN\ QUAD} = \begin{pmatrix} 1 & 0 & 0 & 0 \\ -1/f & 1 & 0 & 0 \\ 0 & 0 & 1 & 0 \\ 0 & 0 & 1/f & 1 \end{pmatrix} \tag{2.26}$$

showing that a quadrupole that focuses in the horizontal plane defocuses in the vertical! Similarly, a horizontally focusing quadrupole of arbitrary length is represented by

$$M_{QUAD} = \begin{pmatrix} \cos(kL) & \frac{1}{k}\sin(kL) & 0 & 0 \\ -k\sin(kL) & \cos(kL) & 0 & 0 \\ 0 & 0 & \cosh(kL) & \frac{1}{k}\sinh(kL) \\ 0 & 0 & k\sinh(kL) & \cosh(kL) \end{pmatrix} \tag{2.27}$$

where

$$k = \sqrt{K} \tag{2.28}$$

and K are both positive. The case of a horizontally defocusing (and vertically focusing) thick quadrupole is discussed in the Appendix.

All the pieces are now in place for a discussion of linear motion through the optical lattice that makes up a closed ring. How is it possible to guarantee that the linear motion is stable? How to deal with the annoying fact that quadrupoles do not simultaneously focus in both planes? How is it possible to manipulate the transverse beam sizes to be small in some places but large in others? Why is it necessary to introduce nonlinear magnets, such as sextupoles?

There are other linear elements besides drifts, dipole and quadrupoles in the accelerator zoo, which also have linear matrices – for example, combined function magnets (with both dipole and quadrupole field components), non-rectangular dipoles, solenoids, skew quadrupoles (rotated by 45 degrees about the longitudinal axis) and accelerating RF sections. Six-by-six matrices for these standard elements are found in the Appendix.

Exercises

2.1 Is the particle shown in Figure 2.5 positively or negatively charged? And in Figure 2.6? Explain your reasoning.

2.2 Derive Equation 2.13 for a singly charged particle, from Equation 2.12. What is the magnetic rigidity of a fully stripped ion with atomic number Z and atomic weight A, when the momentum is measured in units of (GeV/u)/c?

2.3 RHIC collides fully stripped gold ions with $(Z,A) = (79, 196.97)$ at a top energy of 100 GeV/nucleon in each beam, in rings of circumference 3834 m. The atomic mass unit is $m_0 = 0.93113$ GeV/c^2.

a) The injection energy is 10.5 GeV/nucleon. What is the required swing in revolution frequency during RHIC acceleration?

b) There are 192 main arc dipoles per ring, with a magnetic length $L =$ 9.42 m. In the approximation that only they provide all 2π radians of bending, what is the dipole field at top energy?

2.4 Derive Equation 2.27 and show that it reduces to Equation 2.26 in the limit that $L \to 0$ while KL is held constant.

2.5 (See also Exercise 5.5.) You inherit a set of quadrupoles $L = 0.5$ m long from a defunct accelerator. The quadrupoles have a 70 mm bore radius and a maximum pole tip field of 1.1 T.

a) What is the minimum focal length f, for a 15 GeV/c beam?

b) Is this quadrupole thin, meaning $L \ll |f|$?

c) How long can a quadrupole with the same cross section extend before it becomes thick, as a function of beam energy?

2.6 Derive the matrix that represents a thin quadrupole with a focal length f that is rotated by 45 degrees about the longitudinal axis – a thin skew quadrupole.

2.7 How is the matrix representing a linear magnet transformed when it is expressed in a co-ordinate frame where X replaces x and $X = -x$?

2.8 Figure 2.8 illustrates an early reel-to-reel tape drive, reading and writing data through an input/output channel to a nearby computer. The fluctuating data transfer rate determines the speed at which tape is fed through the

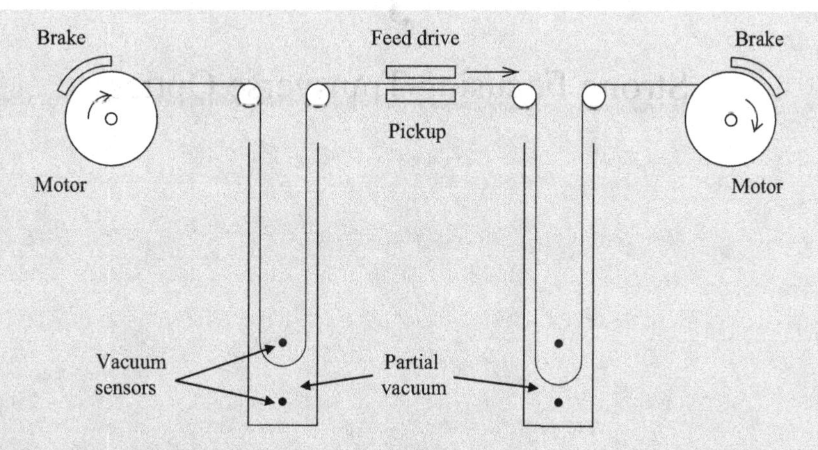

Figure 2.10 *An early computer reel-to-reel tape drive. The brake, the motor or neither is applied to each reel, depending on the length of tape in the nearby vacuum buffer.*

magnetic pickup. The amount of loose tape sucked into nearby partial-vacuum buffers is sensed, turning on or off the motor or brake on the nearby reel, or allowing the reel to freewheel. Assume that the sensors are perfect, and that the reel motor acceleration and brake deceleration rates are simple constants.

a) Is the system prone to uncontrolled oscillations that could break the tape? Under what conditions?
b) Clearly, this dynamical system is not readily amenable to representation by differential equations. Although numerical simulation is relatively straightforward, might it introduce artefacts that are not present in the real world?
c) Illustrate your answers with results generated by a numerical simulation.

3

Strong Focusing Transverse Optics

The matrices representing linear motion through optical elements can be multiplied together to generate a one-turn matrix at a reference point

$$M = M_{m(m-1)} \ \ldots \ M_{21} M_{10} \tag{3.1}$$

where the matrix M_{kj} represents motion from s_j to s_k, as illustrated in Figures 3.1 and 3.2. Is the transverse motion stable? Are the horizontal and vertical phase space vectors (x, x') and (y, y') both bounded as time, the turn number n, goes to infinity?

For simplicity, it is assumed that a bend is a pure dipole in a separated function lattice that contains only drifts, dipoles and quadrupoles. (Nonlinear magnets, such as sextupoles, octupoles and decapoles, are added later.) Nonetheless, the same mathematical framework developed below works with combined function bends, which have superimposed dipole and quadrupole fields. Combined function magnets have been used from the earliest days of strong focusing accelerators, for example, in the venerable Alternating Gradient Synchrotron (AGS) at Brookhaven National Laboratory (BNL), and the Proton Synchrotron (PS) at CERN. They have some compactness advantages, especially for medium- and low-energy hadron rings.

Assuming that horizontal and vertical motion is decoupled so that the M_{kj} are all block diagonal, then horizontal stability concerns the 2×2 matrix equation

$$\vec{x}_n = M^n \, \vec{x}_0 \tag{3.2}$$

where the horizontal phase space vector

$$\vec{x} = \begin{pmatrix} x \\ x' \end{pmatrix} \tag{3.3}$$

has the initial value $\vec{x}_0$. Vertical stability concerns a similar, independent equation. Even in the idealised assumption of perfect decoupling, there is still a fundamental

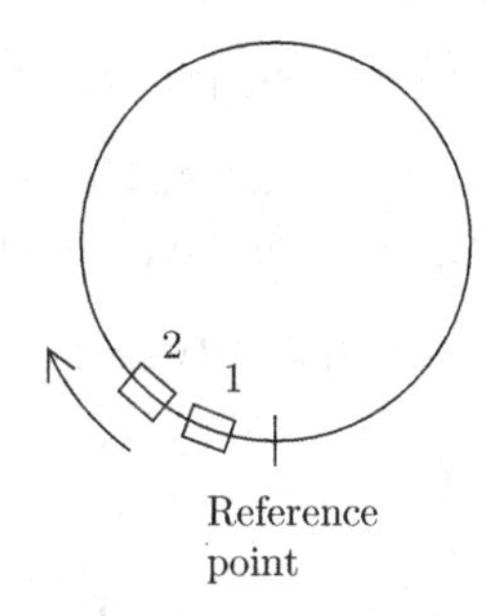

Figure 3.1 The matrices that represent optical elements in the lattice of an accelerator ring are multiplied together in sequence to derive the one-turn matrix at a convenient reference point.

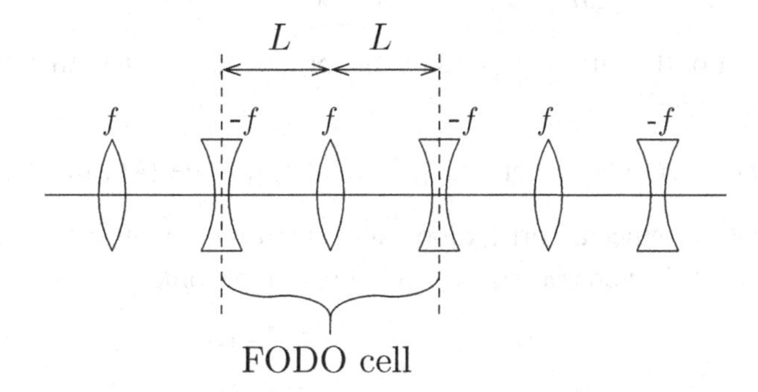

Figure 3.2 The lattice of an accelerator often includes a basic cell that is repeated many times. For example, a separated function FODO cell consists of uniformly spaced alternating gradient quadrupoles. The distance L between the quadrupoles contains a drift or a dipole bending magnet, or a combination of both.

problem: a horizontally focusing quadrupole defocuses in the vertical plane, and vice versa. How to focus in both planes simultaneously? The solution lies in strong focusing.

3.1 Linear Stability and Twiss Functions

The one-turn 2×2 matrix M has two complex eigenvectors ($\vec{v}_1$ and $\vec{v}_2$) and two complex eigenvalues (λ_1 and λ_2) that solve the equation

$$M\vec{v} = \lambda\vec{v} \tag{3.4}$$

If the initial vector is written as

$$\vec{x}_0 = A\vec{v}_1 + B\vec{v}_2 \tag{3.5}$$

where both sides of the equation are real (although constants A and B are complex), then on turn n the phase space vector is just

$$\vec{x}_n = M^n \vec{x}_0 = A\lambda_1^n \vec{x}_1 + B\lambda_2^n \vec{x}_2 \tag{3.6}$$

Clearly, if $\vec{x}_n$ is to be bounded for all n, then so also must λ_1^n and λ_2^n.

The eigenvalues λ are found by solving the characteristic equation

$$\det(M - \lambda I) = 0 \tag{3.7}$$

where I is the unit 2×2 matrix. If

$$M = \begin{pmatrix} a & b \\ c & d \end{pmatrix} \tag{3.8}$$

then the characteristic equation becomes the quadratic equation

$$(ad - bc) - (a + d)\lambda + \lambda^2 = 0 \tag{3.9}$$

The determinant of the one-turn matrix is the product of the determinants of all the matrices in the lattice,

$$\det(M) = ad - bc = \det(M_{m(m-1)}) \ldots \det(M_{21}) \det(M_{10}) = 1 \tag{3.10}$$

and has the value 1, since all drift, dipole and quadrupole matrices are unimodular, with $\det(M_{kj}) = 1$. The characteristic equation then becomes

$$1 - (a + d)\lambda + \lambda^2 = 0 \tag{3.11}$$

which is trivially rewritten in terms of the trace $\mathrm{Tr}(M)$

$$\lambda^{-1} + \lambda = \mathrm{Tr}(M) = a + d \tag{3.12}$$

where the sums on both sides of the equation are real, since M is real.

It is seen by inspection that the eigenvalues are reciprocals so that

$$\lambda_1 = e^{i\mu}, \qquad \lambda_2 = e^{-i\mu} \tag{3.13}$$

where μ, which may be complex, is found by solving the equation

$$2\cos(\mu) = \mathrm{Tr}(M) \tag{3.14}$$

If μ *is* complex, then either λ_1^n or λ_2^n increases without bound as $n \to \infty$, and the linear motion is unstable. Therefore, the condition for linear stability – for μ to be real – is

$$-1 \le \frac{1}{2}\mathrm{Tr}(M) \le 1 \tag{3.15}$$

If the motion is stable, then the one-turn matrix is written in general as

$$M(s) = \begin{pmatrix} \cos(\mu) + \alpha \sin(\mu) & \beta \sin(\mu) \\ -\gamma \sin(\mu) & \cos(\mu) - \alpha \sin(\mu) \end{pmatrix} \tag{3.16}$$

where $\alpha(s), \beta(s)$ and $\gamma(s)$ – the Courant–Snyder or Twiss functions – satisfy the relationship

$$\gamma \equiv \frac{1+\alpha^2}{\beta} \tag{3.17}$$

thereby guaranteeing that $M(s)$ is unimodular. Note that β has the dimension of length, while α is dimensionless. The Twiss functions vary with s as the reference point is moved around the ring. In contrast, the trace of the matrix is invariant so that μ is independent of location: it is a property of the lattice as a whole.

3.2 Turn-by-Turn Motion in Phase Space

The one-turn matrix of Equation 3.16 is written more succinctly as

$$M = T^{-1} R T \tag{3.18}$$

where the co-ordinate rotation matrix

$$R(\mu) = \begin{pmatrix} \cos(\mu) & \sin(\mu) \\ -\sin(\mu) & \cos(\mu) \end{pmatrix} \tag{3.19}$$

has the property

$$R^n(\mu) = R(n\mu) \tag{3.20}$$

and the Floquet transformation matrix

$$T = \begin{pmatrix} 1/\sqrt{\beta} & 0 \\ \alpha/\sqrt{\beta} & \sqrt{\beta} \end{pmatrix} \tag{3.21}$$

has the inverse

$$T^{-1} = \begin{pmatrix} \sqrt{\beta} & 0 \\ -\alpha/\sqrt{\beta} & 1/\sqrt{\beta} \end{pmatrix} \tag{3.22}$$

The phase space vector after n turns is therefore simply

$$\begin{aligned} \vec{x}_n &= M^n \, \vec{x}_0 \\ &= (T^{-1}RT)\,(T^{-1}RT)\,\ldots\,(T^{-1}RT)\,\vec{x}_0 \\ &= T^{-1}R^n\,T\,\vec{x}_0 \end{aligned} \tag{3.23}$$

Turn-by-turn motion is simplified even further by introducing normalised phase space co-ordinates $(\tilde{x}, \tilde{x}')$, related to the physical co-ordinates (x, x') through the Floquet transformation

$$\vec{\tilde{x}} = T\vec{x} \tag{3.24}$$

because then

$$\vec{\tilde{x}}_n = R(n\mu)\, \vec{\tilde{x}}_0 \tag{3.25}$$

with the trivial solution

$$\begin{aligned} \tilde{x}_n &= a \sin(2\pi Q_x n + \phi_0) \\ \tilde{x}'_n &= a \cos(2\pi Q_x n + \phi_0) \end{aligned} \tag{3.26}$$

where the constants a and ϕ_0 are the initial amplitude and phase, respectively. This motion, on a circle of radius a, is illustrated in Figure 3.3a.

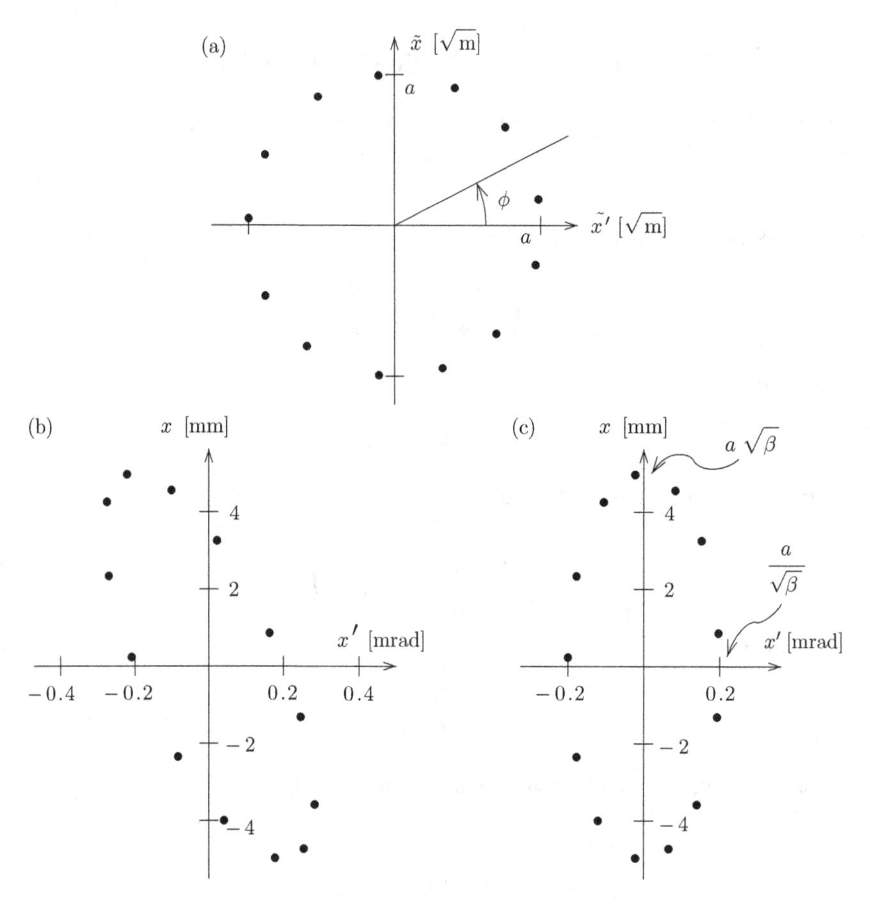

Figure 3.3 Linear motion for 12 turns with a fractional tune of mod $(Q_x, 1) = 0.155$. The motion in normalised phase space, shown in (a), is on a circle of radius a, advancing by an angle of $\Delta\phi = 2\pi Q_x$ on every turn. The same motion is shown in physical phase space in (b), when $a = 0.001\ m^{1/2}$ at a reference point with $\alpha = 1.0$ and $\beta = 25$ m. Motion is on a tilted ellipse, with extreme displacements of $x = \pm a\sqrt{\beta} = 5$ mm. Motion in physical phase space at another reference point with $\alpha = 0$ is shown in (c). Motion is on an erect ellipse, with an aspect ratio of $\beta = 25$ m.

The introduction of the betatron tune Q_x, through

$$\mu = 2\pi Q_x \tag{3.27}$$

underlines the connection between Equations 2.1 and 3.26, although the dimensions of x (length) and $\tilde{x}$ (square root of length) are not the same. Inverting Equation 3.24 and using Equation 3.22 leads to the turn-by-turn solution in physical phase space

$$\begin{aligned} x_n &= a\sqrt{\beta}\ \sin(2\pi Q_x n + \phi_0) \\ x_n' &= a/\sqrt{\beta}\ [\cos(2\pi Q_x n + \phi_0) - \alpha \sin(2\pi Q_x n + \phi_0)] \end{aligned} \tag{3.28}$$

Figures 3.3b and 3.3c illustrate this motion at two reference points, both with a typical value of $\beta = 25$ m. Motion is in general on a tilted ellipse, unless $\alpha = 0$, in which case the aspect ratio of the erect ellipse is β.

3.3 Propagation across a Fraction of a Turn

Motion in physical phase space from s_1 to s_2 – between any two locations within a single turn – is written as

$$\vec{x}_2 = M_{21}\vec{x}_1 \tag{3.29}$$

where the transfer matrix

$$M_{21} = T_2^{-1}\, R(\phi_2 - \phi_1)\, T_1 \tag{3.30}$$

consists of a transformation into normalised phase space at s_1, followed by a rotation, and a transformation back to physical phase space at s_2. Written out explicitly, this becomes

$$M_{21} = \begin{pmatrix} \sqrt{\frac{\beta_2}{\beta_1}}\,(c_{21} + \alpha_1 s_{21}) & \sqrt{\beta_2\beta_1}\, s_{21} \\ \frac{-(1+\alpha_1\alpha_2)\, s_{21} + (\alpha_2 - \alpha_1)\, c_{21}}{\sqrt{\beta_2\beta_1}} & \sqrt{\frac{\beta_1}{\beta_2}}\,(c_{21} - \alpha_2 s_{21}) \end{pmatrix} \tag{3.31}$$

where, for brevity

$$\begin{aligned} s_{21} &= \sin(\phi_2 - \phi_1) \\ c_{21} &= \cos(\phi_2 - \phi_1) \end{aligned} \tag{3.32}$$

Equation 3.31 reverts to the much simpler form of Equation 3.16 in the case when $s_2 = s_1 + C$, one turn later.

Twiss functions and betatron phases can also be propagated from s_1 to s_2, given only a knowledge of the element-by-element transfer matrices in between. If the transfer matrix is written for brevity as

$$M_{21} = \begin{pmatrix} m_{11} & m_{12} \\ m_{21} & m_{22} \end{pmatrix} \tag{3.33}$$

then it can be shown that Twiss functions propagate through

$$\begin{pmatrix} \beta_2 \\ \alpha_2 \\ \gamma_2 \end{pmatrix} = \begin{pmatrix} m_{11}^2 & -2m_{11}m_{12} & m_{12}^2 \\ -m_{21}m_{11} & 1+2m_{12}m_{21} & -m_{12}m_{22} \\ m_{21}^2 & -2m_{22}m_{21} & m_{22}^2 \end{pmatrix} \begin{pmatrix} \beta_1 \\ \alpha_1 \\ \gamma_1 \end{pmatrix} \tag{3.34}$$

The phase advance from s_1 to s_2 is found by inverting the equation

$$\tan(\phi_2 - \phi_1) = \frac{m_{12}}{m_{11}\beta_1 - m_{12}\alpha_1} \tag{3.35}$$

giving an unambiguous result if the phase advance is not too large.

3.4 Continuous Propagation

Sometimes, it is convenient or necessary to consider transverse motion as a continuous function of s. Difference and differential formalisms for transverse motion are connected by considering incremental motion. For a small enough distance Δs

$$\begin{aligned} c_{21} &\approx 1 \\ s_{21} &\approx \Delta\phi \end{aligned} \tag{3.36}$$

and the transfer matrix of Equation 3.31 becomes

$$M_{21} \approx \begin{pmatrix} \left(1 + \frac{1}{2}\frac{\Delta\beta}{\beta}\right)(1 + \alpha\Delta\phi) & \beta\,\Delta\phi \\ \sim & \left(1 - \frac{1}{2}\frac{\Delta\beta}{\beta}\right)(1 - \alpha\Delta\phi) \end{pmatrix} \tag{3.37}$$

where the (unexpanded) bottom-left matrix element ensures that M_{21} is unimodular. This is a drift matrix in the limit that $\Delta s \to 0$

$$M_{21} \approx \begin{pmatrix} 1 & \Delta s \\ \sim & 1 \end{pmatrix} \tag{3.38}$$

so it is immediately clear that

$$\frac{d\phi}{ds} = \frac{1}{\beta} \tag{3.39}$$

$$\alpha = -\frac{1}{2}\frac{d\beta}{ds} \tag{3.40}$$

These differential equations are the first to appear in this chapter. They are derived from difference equations, not *vice versa!*

Equation 3.39 shows that phase advances can be calculated, formally speaking, by continuous integration, since

$$\phi(s_2) - \phi(s_1) = \int_1^2 \frac{ds}{\beta} \tag{3.41}$$

Similarly, the betatron tune Q_x can be calculated by a complete integration around the ring, since

$$2\pi\, Q_x = \oint \frac{ds}{\beta} \tag{3.42}$$

These two integral forms are sometimes convenient analytically. However, practical calculations of tunes and phase advances rely on the application of Equations 3.14 and 3.35.

Continuous transverse motion around a fraction of a turn is written as

$$x(s) = a\sqrt{\beta(s)}\, \sin(\phi(s) + \phi_0) \tag{3.43}$$

Direct differentiation with respect to s, using Equations 3.39 and 3.40, gives

$$x'(s) = \frac{dx}{ds} = a/\sqrt{\beta(s)}\, [\cos(\phi(s) + \phi_0) - \alpha(s) \sin(\phi(s) + \phi_0)] \tag{3.44}$$

Equations 3.43 and 3.44 naturally reduce to Equation 3.28, for turn-by-turn motion.

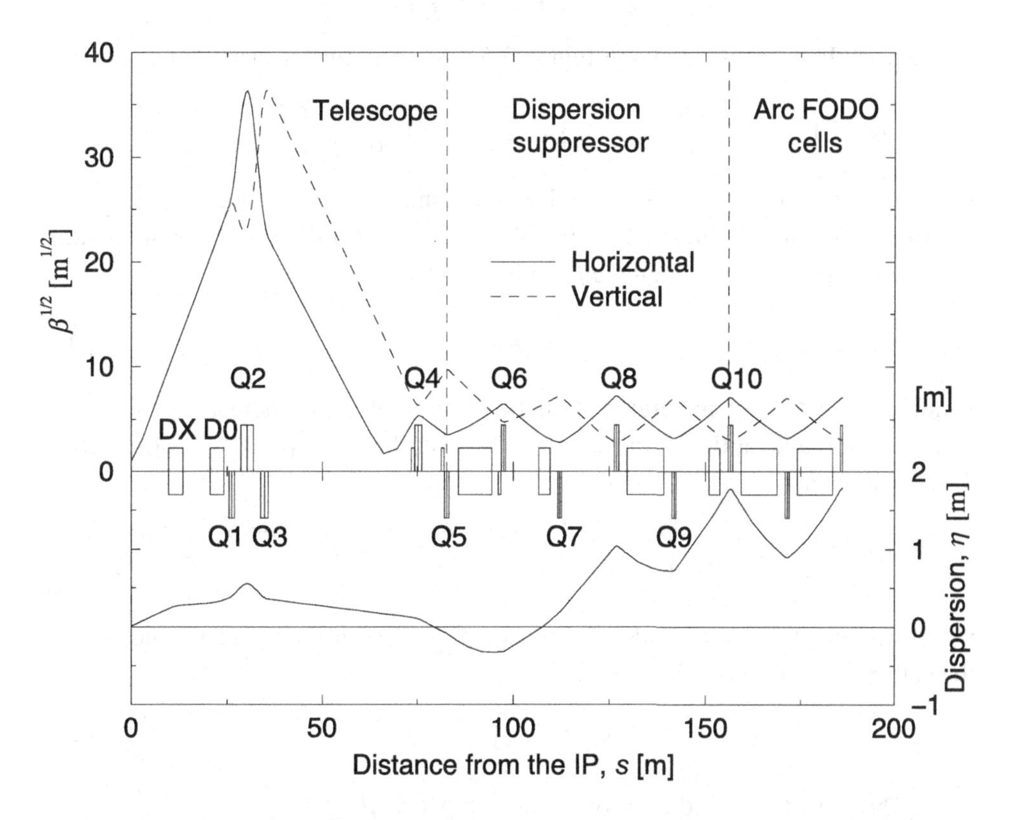

Figure 3.4 Modular manipulation of optics near a RHIC interaction point (IP), showing the optical behaviour of the horizontal and vertical Twiss β-functions, and the dispersion function η, as they propagate through a telescope and a dispersion suppressor. Dipoles are shown as long rectangles centred on the horizontal axis. Horizontally focusing or defocusing quadrupoles are shown as short rectangles above or below the axis [18].

A practical example of accelerator optics is the first 200 m downstream of a RHIC interaction point, where counter-rotating beams collide [18]. Figure 3.4 shows how the horizontal and vertical Twiss β-functions are first manipulated by adjusting the strengths of quadrupoles Q1, Q2, Q3 and Q4 in the telescope section, constraining them to have values $\beta < 80$ m when $s > 70$ m. Next, in the dispersion suppressor, a combination of regularly spaced but independently powered quadrupoles (Q5 through to Q10) and irregularly spaced dipoles, shown as rectangular boxes, manipulates the dispersion function to have a value $\eta \approx 1.8$ m and a slope $d\eta/ds = 0$ at the entrance to the arc FODO cells. (Dispersion functions, which describe off-momentum optics, are discussed in Chapter 4). The horizontal and vertical β-functions and the dispersion function oscillate between fixed 'matched' values in the many identical FODO cells in each of the 6 main RHIC arcs.

3.5 FODO Cell Optics

The generic FODO cell shown in Figure 3.5 has a net transfer matrix of

$$M_{FODO} = \begin{pmatrix} 1 & 0 \\ -q & 1 \end{pmatrix} \begin{pmatrix} 1 & L \\ 0 & 1 \end{pmatrix} \begin{pmatrix} 1 & 0 \\ 2q & 1 \end{pmatrix} \begin{pmatrix} 1 & L \\ 0 & 1 \end{pmatrix} \begin{pmatrix} 1 & 0 \\ -q & 1 \end{pmatrix} \tag{3.45}$$

where the reference point is located at the centre of a focusing F quadrupole. A distance L separates negligibly thin F and D quadrupoles that have alternating equal and opposite focal lengths $\pm f$, with a positive value

$$q = \frac{1}{2f} \tag{3.46}$$

which represents their integrated strength. If the FODO cell is matched so that β and α are the same at the entrance and at the exit, then the application of Equation 3.31 shows that

$$M_{FODO} = \begin{pmatrix} 1 - 2(qL)^2 & 2L(1 + (qL)) \\ -2q(qL)(1 - qL) & 1 - 2(qL)^2 \end{pmatrix} = \begin{pmatrix} C + \alpha S & \beta S \\ -\gamma S & C - \alpha S \end{pmatrix} \tag{3.47}$$

where, for brevity, $S = \sin(\Delta\phi)$, $C = \cos(\Delta\phi)$ and $\Delta\phi$ is the phase advance across the cell. Solving Equation 3.47 gives

$$\sin\left(\frac{\Delta\phi}{2}\right) = |qL| \tag{3.48}$$

which shows that the condition for linear stability is $qL \leq 1$, or

$$f \geq \frac{L}{2} \tag{3.49}$$

Stability is only possible if the quadrupoles are weak enough!

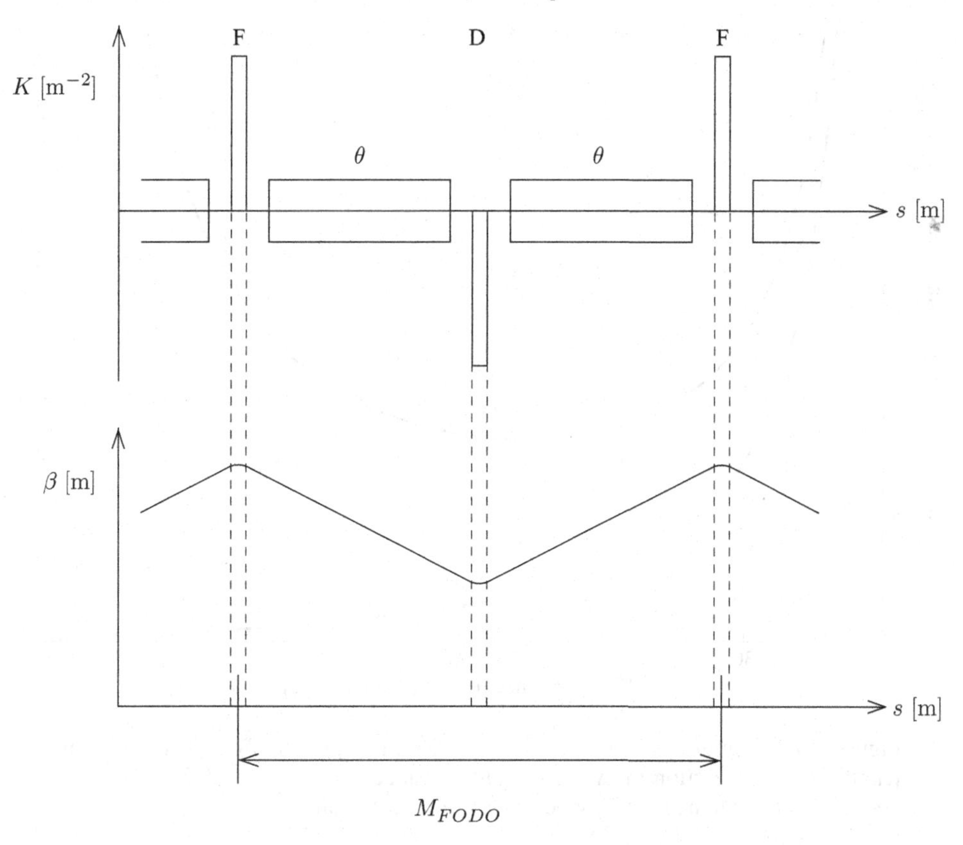

Figure 3.5 A generic FODO cell with equally spaced quadrupoles of equal and opposite strength, containing dipoles with a bend angle θ. The matched β-function is exactly periodic, over a total cell length of $2L$.

Solving Equation 3.47, and a similar equation for M_{DOFO} at the centre of a D quadrupole, shows that the maximum and minimum β-values are

$$\frac{\beta_{max}}{L} = \left(\frac{1+s}{sc}\right) = \frac{1}{s}\left(\frac{1+s}{1-s}\right)^{1/2} \tag{3.50}$$
$$\frac{\beta_{min}}{L} = \left(\frac{1-s}{sc}\right) = \frac{1}{s}\left(\frac{1-s}{1+s}\right)^{1/2}$$

where it is convenient to introduce the (potentially confusing) notation

$$s = \sin(\Delta\phi/2) \tag{3.51}$$
$$c = \cos(\Delta\phi/2)$$

Figure 3.6 shows that β_{max} has a broad minimum (at constant L) between about 60 and 90 degrees, a range that is commonly used in practice.

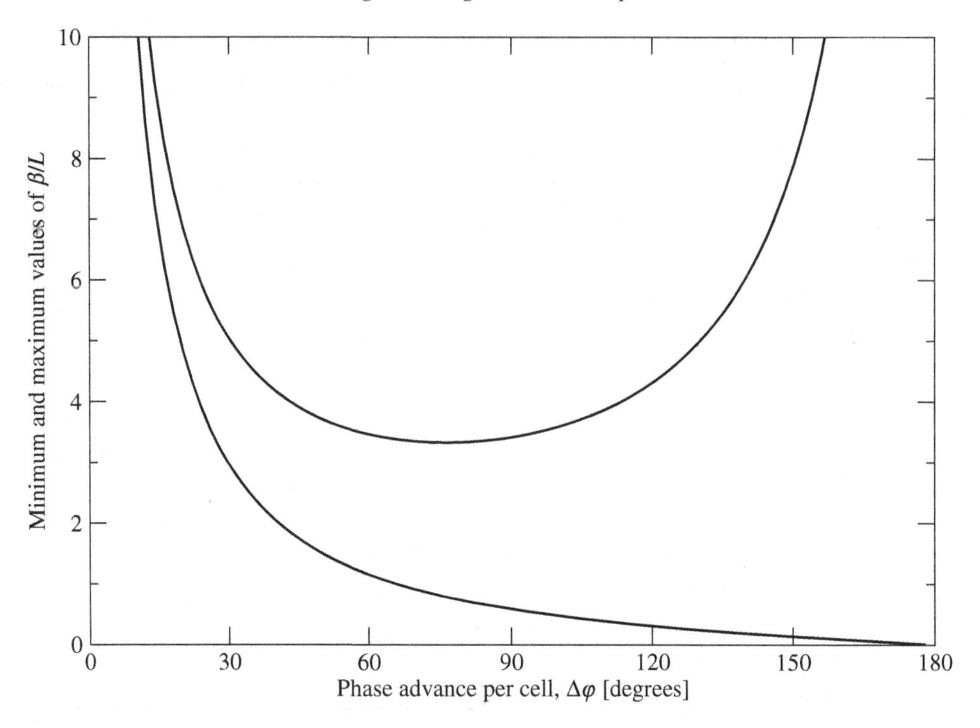

Figure 3.6 Minimum and maximum β-functions in a generic FODO cell of total length $2L$, as a function of $\Delta\phi$, the phase advance per cell. The β-functions are assumed to be matched, exactly periodic from cell to cell.

A simple way to adjust typical β-function values is to change the cell length $2L$, since β scales like L. There is a practical limit to β-function reduction, however, because the integrated quadrupole strength increases as L is decreased, and the quadrupoles become longer. At some point, the quadrupoles are no longer thin, and thus, the dipole packing fraction is significantly reduced. There is a practical minimum to the FODO cell length.

So far, it has been assumed that thin quadrupoles are evenly spaced, that they are of equal and opposite strengths, and that the dipole bend angle θ is small so that pole-face rotation effects are insignificant. In such a FODO channel, the horizontal and vertical β-function minima and maxima are identical, except that their locations alternate, as illustrated in Figure 3.4.

In practice, the F and D quadrupoles often have somewhat different absolute strengths, splitting the horizontal and vertical phase advances per cell. Similarly, the F and D quadrupoles in linacs (linear accelerators) are often placed asymmetrically in a doublet configuration, with one long and one short drift per cell. Nonetheless, the general features of the FODO cell persist. Most fundamentally, if all the lengths of any repetitive cell structure are doubled, the matched β-functions also double in size.

Exercises

3.1 Consider the one-turn matrix M given in Equation 3.16.

a) Show that M is unimodular.
b) Show that the trace of M does not vary as the reference point moves around the lattice from s_1 to s_2. Hint: pre- and post-multiply M by M_{21} and M_{21}^{-1}, where M_{21} represents motion from s_1 to s_2.

3.2 What are the extreme values of the physical angle x', at a reference point with arbitrary β and α values, for betatron motion that has an amplitude of a in normalised phase space?

3.3 Prove that $M = T^{-1}RT$, where M is a 2×2 one-turn matrix, R is a rotation matrix and T is the Floquet transformation matrix defined in Equation 3.21.

3.4 Prove Equations 3.34 and 3.35.

3.5 Consider a long field-free region adjacent to a collision point.

a) Use Equation 3.34 to show that the β-function evolves with s like

$$\beta(s) = \beta^* + \frac{(s - s_0)^2}{\beta^*} \tag{3.52}$$

b) How does $\alpha(s)$ evolve?
c) How does the phase $\phi(s)$ evolve?
d) What is the largest phase advance possible, across a field-free region?
e) Why are optics often displayed as $\sqrt{\beta}$ vs. s, rather than β vs. s?

3.6 Consider the net effect of axially symmetric convex and concave lens on a light beam. If two lenses with focal lengths f and $-f$ are placed distance L apart, demonstrate that there is net focussing if $|f| > L$, no matter which way round the lenses are placed.

3.7 Consider the FODO cell matrix

$$M = \begin{pmatrix} 1 - 2(qL)^2 & 2L(1 + (qL)) \\ -2q(qL)(1 - qL) & 1 - 2(qL)^2 \end{pmatrix} \tag{3.53}$$

a) Demonstrate directly that M is unimodular.
b) What are the eigenvalues and eigenvectors of M? Why doesn't it matter that they are imaginary?

3.8 A particle initially following an ideal trajectory is deflected horizontally by a perturbation angle θ at a location where the beta function is β_0.

a) Show that the motion downstream of this perturbation is given by

$$x(s) = \theta \sqrt{\beta_0 \beta(s)} \sin\phi \tag{3.54}$$

where $\phi = \phi(s) - \phi_0$ is the phase advance to the observation point.

b) How big in radians can the deflection θ be without the particle hitting the beampipe? (State your assumptions about typical values of β and beampipe size.)

3.9 A sequence of magnets transports beam from one location to another. The net motion is described by a 'black box' matrix

$$M = \begin{pmatrix} a & b \\ c & d \end{pmatrix} \tag{3.55}$$

a) A particle enters the front face of the black box parallel to the optical axis. How far from the exit of the black box does it cross the axis? Define this length to be the focal length of an arbitrarily long and arbitrarily complex sequence of magnets.

b) Consider the black box to be a thick horizontally focusing quadrupole. What are the horizontal and vertical focal lengths?

c) Plot both focal lengths as a function of the quadrupole length.

4

Longitudinal and Off-Momentum Motion

So far we have considered – almost exclusively in a matrix formalism – the transverse motion of on-momentum test particles about a closed orbit that replicates itself, turn-by-turn. In the absence of errors the closed orbit is the same as the design orbit, usually down the centre of the beampipe. Now it is time to consider the motion of off-momentum particles, first discussing the transverse motion of constant momentum particles, and then the longitudinal motion of particles with oscillating momentum deviations. It is convenient to do this by reverting to a differential formalism that begins with Hill's equations of on-momentum transverse motion

$$\begin{aligned} x'' + K(s)\,x &= 0 \\ y'' - K(s)\,y &= 0 \end{aligned} \tag{4.1}$$

where the local quadrupole strength K is a periodic function in the azimuthal coordinate s, and a prime indicates differentiation with respect to s.

4.1 Constant Momentum Offset: Transverse Motion

A particle with momentum p that deviates from the nominal p_0 by

$$\delta = \frac{p - p_0}{p_0} = \frac{\Delta p}{p} \tag{4.2}$$

has a radius R when passing through a dipole, according to

$$\begin{aligned} \frac{1}{R} &= \frac{1}{\rho}\,\frac{1}{(1+\delta)} \\ &= G\,\frac{1}{(1+\delta)} \end{aligned} \tag{4.3}$$

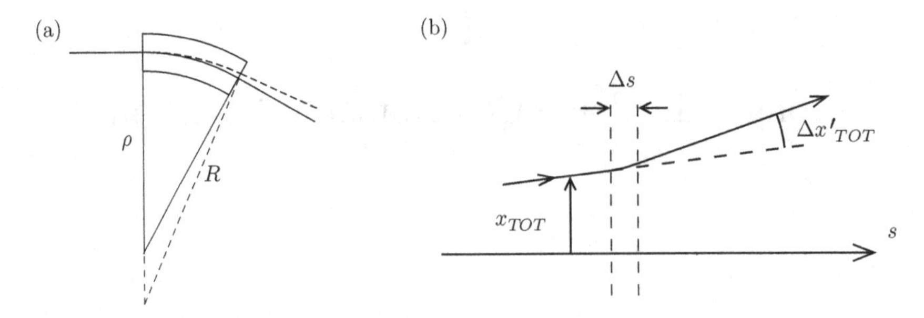

Figure 4.1 Motion of an off-momentum particle through a dipole. The plan view (a) shows a particle with a momentum p that is slightly larger than the nominal p_0, with a bending radius R slightly larger than the nominal value ρ. In the curvilinear co-ordinate frame (b) the particle acquires a horizontal angle $\Delta x'_{TOT}$ when passing through a thin slice of length Δs.

as illustrated in Figure 4.1a. It acquires an additional angle $\Delta x'_{TOT}$ when passing through a thin dipole slice of length Δs, where

$$\Delta x'_{TOT} = G\left(1 - \frac{1}{1+\delta}\right) \Delta s \tag{4.4}$$

The subscript *TOT* is introduced here, in anticipation of the separation of the total displacement (and angle) into two pieces – one for the offset of the closed orbit due to the off-momentum parameter δ, and the other for free betatron oscillations around that offset closed orbit.

The differential equation for horizontal motion through a dipole is now

$$x''_{TOT} = G\left(1 - \frac{1}{1+\delta}\right) \tag{4.5}$$

and for off-momentum horizontal motion through a quadrupole is

$$x''_{TOT} + \frac{K}{(1+\delta)} x_{TOT} = 0 \tag{4.6}$$

so that the generalised version of Hill's equations is

$$\begin{aligned} x''_{TOT} + \frac{K}{(1+\delta)} x_{TOT} &= G\left(1 - \frac{1}{1+\delta}\right) \\ y''_{TOT} - \frac{K}{(1+\delta)} y_{TOT} &= 0 \end{aligned} \tag{4.7}$$

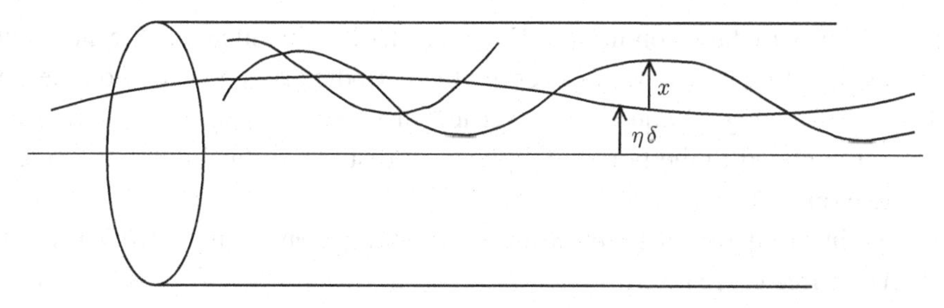

Figure 4.2 The total horizontal displacement of an off-momentum particle consists of the constant closed orbit offset $\eta\delta$, plus a free betatron oscillation of displacement x that varies from turn to turn.

where it is assumed that the dipoles bend only in the horizontal plane. These equations are exact – they are correct to all orders in δ – but often the first order approximation is sufficient:

$$\begin{aligned} x''_{TOT} + K(1-\delta)\, x_{TOT} &\approx G\,\delta \\ y''_{TOT} - K(1-\delta)\, y_{TOT} &\approx 0 \end{aligned} \tag{4.8}$$

4.2 The Dispersion Function

The total horizontal displacement of a test particle is

$$x_{TOT} = \eta(s)\,\delta + x \tag{4.9}$$

as illustrated in Figure 4.2. The dispersion function $\eta(s)$ varies with s around the accelerator, but does not depend on turn number. In contrast, the free oscillation displacement x (relative to the offset closed orbit) does vary with turn number. Setting $x = 0$ and substituting Equation 4.9 into Equation 4.7 gives

$$\eta'' + \frac{K}{(1+\delta)}\,\eta = \frac{G}{\delta}\left(1 - \frac{1}{1+\delta}\right) \tag{4.10}$$

which is the exact equation that is to be solved to find η for a particular value of δ, after the application of periodic boundary conditions in s.

Strictly speaking the dispersion function is a polynomial in δ

$$\eta(s) = \eta_0(s) + \eta_1(s)\,\delta + \eta_2(s)\,\delta^2 + \ldots \tag{4.11}$$

but it is common to consider only the first term in the expansion, an excellent approximation in large accelerators. In this case the dispersion function is

$$\eta'' + K\,\eta = G \tag{4.12}$$

with periodic boundary conditions. Equation 4.12 is identical to the horizontal Hill's equation, Equation 4.1, after adding the periodic driving term G on the right hand side. The analogous equation for vertical dispersion has no driving term for an accelerator confined to the horizontal plane, so that the vertical dispersion is zero everywhere (except for errors).

Multiplying Equation 4.10 by δ and subtracting from Equation 4.7 gives the equations for free oscillations

$$\begin{aligned} x'' + \frac{K}{(1+\delta)} x = 0 \\ y'' - \frac{K}{(1+\delta)} y = 0 \end{aligned} \tag{4.13}$$

These exact equations show how the quadrupole geometric strength changes as the momentum varies, inevitably causing the horizontal and vertical β-functions and the betatron tunes Q_x and Q_y to vary with momentum. Tune variations with momentum – the horizontal and vertical chromaticities – become large in large accelerators, and must be suppressed, making it necessary to introduce nonlinear sextupole magnets. Ironically, such nonlinear magnets drive dangerous resonant behaviour, making it doubly necessary to control the chromaticities and the tunes.

Figure 3.4 shows how the dispersion function evolves from a value of (practically) zero at one of RHIC's interaction points, into well-behaved values in the arc. The dispersion suppressor region matches the relatively small dispersion values that exit the end of the telescope into the repetitive arc FODO cell values. This is done in part by controlling the driving term G (by removing some of the dipoles), and in part by adjusting the geometric strengths K in quadrupoles Q5 to Q10. The missing dipole locations are convenient for installing specialised items, such as injection kickers and septum magnets.

If the bending radius ρ is constant in a dipole of length $L = \rho\theta$, then integrating Equation 4.12 (with $K = 0$) confirms that the dispersion and its slope at the dipole exit are related to the values at the entrance through

$$\begin{pmatrix} \eta \\ \eta' \\ 1 \end{pmatrix}_{out} = \begin{pmatrix} 1 & L & \rho(1-\cos\theta) \\ 0 & 1 & \sin\theta \\ 0 & 0 & 1 \end{pmatrix} \begin{pmatrix} \eta \\ \eta' \\ 1 \end{pmatrix}_{in} \tag{4.14}$$

This can be generalised to 3×3 reductions of all the 6×6 transfer matrices M_{21} listed in the Appendix, so that

$$\begin{pmatrix} \eta \\ \eta' \\ 1 \end{pmatrix}_{out} = \begin{pmatrix} m_{11} & m_{12} & m_{16} \\ m_{21} & m_{22} & m_{26} \\ 0 & 0 & 1 \end{pmatrix} \begin{pmatrix} \eta \\ \eta' \\ 1 \end{pmatrix}_{in} \tag{4.15}$$

where $m_{61} = m_{62} = 0$ and $m_{66} = 1$ for all non-accelerating components. If M is the one turn 3×3 matrix at a reference point, found by multiplying all such component matrices, then the application of periodic boundary conditions is equivalent to asserting

$$\begin{pmatrix} \eta \\ \eta' \\ 1 \end{pmatrix} = M \begin{pmatrix} \eta \\ \eta' \\ 1 \end{pmatrix} \tag{4.16}$$

These linear equations are readily solved for η and η' at that location.

Similarly, matrix methods also conveniently generate the dispersion functions in a FODO cell. For example, the dispersion and its slope evolve from the centre of the thin D quadrupole shown in Figure 3.5 to the centre of the thin F quadrupole according to

$$\begin{pmatrix} \eta \\ \eta' \\ 1 \end{pmatrix}_F = \begin{pmatrix} 1 & 0 & 0 \\ -q & 1 & 0 \\ 0 & 0 & 1 \end{pmatrix} \begin{pmatrix} 1 & L & \frac{1}{2}L\theta \\ 0 & 1 & \theta \\ 0 & 0 & 1 \end{pmatrix} \begin{pmatrix} 1 & 0 & 0 \\ q & 1 & 0 \\ 0 & 0 & 1 \end{pmatrix} \begin{pmatrix} \eta \\ \eta' \\ 1 \end{pmatrix}_D \tag{4.17}$$

where it is assumed that the bend angle θ is small, and that a single dipole fills all the available space L between the quadrupoles. (See Exercise 4.1 for more discussion.) The dispersion function is matched – repeats exactly from cell to cell – if the slopes at F and D are zero. Expanding Equation 4.17 and asserting that $\eta'_F = \eta'_D = 0$ leads to

$$\eta_{max} = L\theta \left(\frac{2+s}{2s^2} \right) \sim \frac{L^2}{\rho} \tag{4.18}$$
$$\eta_{min} = L\theta \left(\frac{2-s}{2s^2} \right) \sim \frac{L^2}{\rho}$$

where, as before,

$$s = \sin(\Delta\phi/2) = |\mathrm{qL}| \tag{4.19}$$

showing that dispersion scales like the square of the cell length L^2 if the bending radius ρ is held fixed. The variation of the minimum and maximum dispersions with phase advance per FODO cell is illustrated in Figure 4.3, complementary to Figure 3.6.

4.3 Oscillating Momentum: Longitudinal Motion

Two factors compete in altering the time it takes for an off-momentum particle to circulate one turn around an accelerator. First, a higher momentum particle goes faster, so it arrives earlier. Second, a higher momentum particle (usually) has a

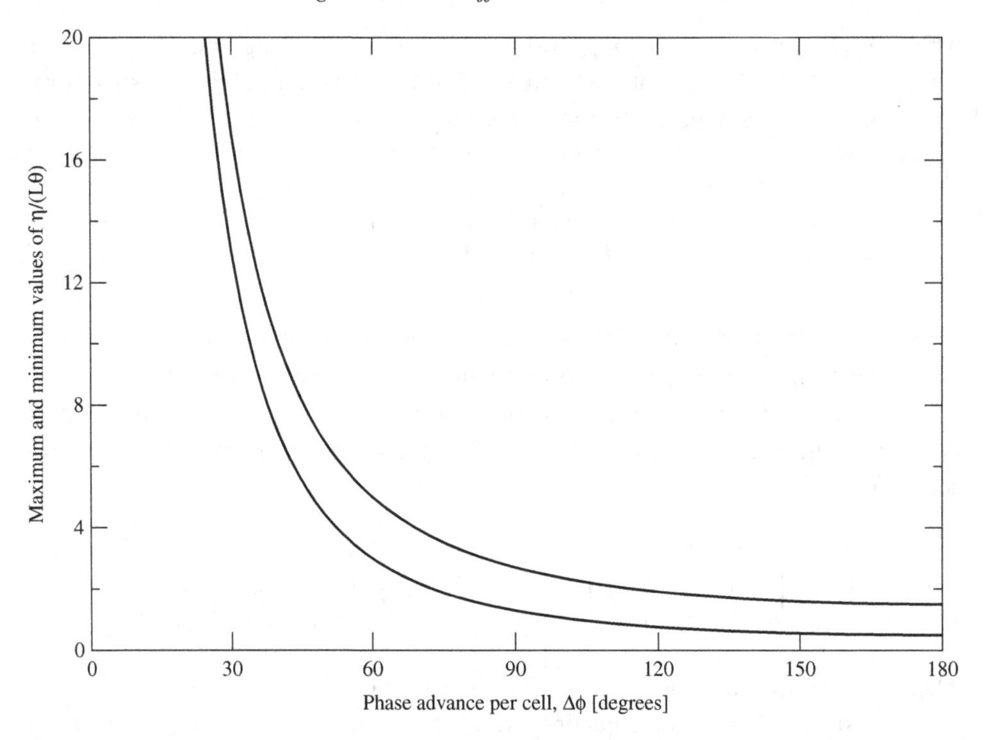

Figure 4.3 Minimum and maximum values of the dispersion function η in a generic FODO cell with a total length $2L$ and a total bend angle of 2θ as a function of $\Delta\phi$, the phase advance per cell. The dispersion function is assumed to be matched, exactly periodic from cell-to-cell.

longer path length, so it tends to arrive later. The first factor is almost irrelevant in the ultra-relativistic limit $\gamma \gg 1$, which is almost always the case for electrons. One or the other factor dominates for protons or heavy ions, depending on whether the nominal γ-value of a bunch is above or below γ_T, the transition gamma or 'energy'.

In one turn, a test particle that is off-momentum by δ has a total additional path length

$$\Delta C_{path} = \delta \oint \eta \, d\theta \tag{4.20}$$

after integration through all the thin angular slices in bending magnets, as illustrated in Figure 4.4. Since

$$\frac{d\theta}{ds} = \frac{1}{\rho} \tag{4.21}$$

then the test particle moves towards the rear of its bunch by

$$\Delta C_{path} = \delta \oint \frac{\eta}{\rho} \, ds \tag{4.22}$$

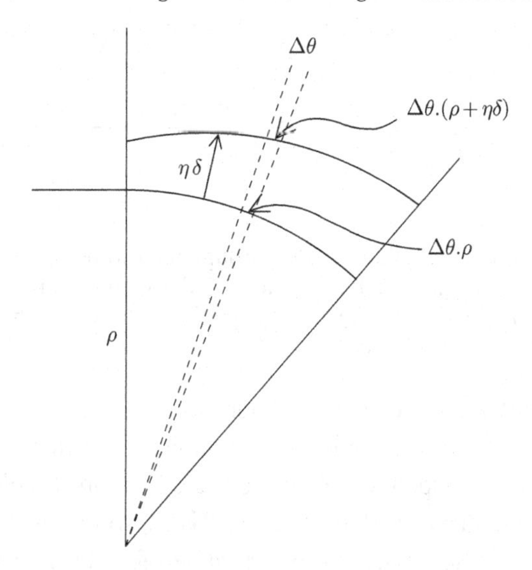

Figure 4.4 An off-momentum particle passing through a thin angular slice $\Delta\theta$ in a bending magnet has an additional path length of $\eta\delta\ \Delta\theta$ when it follows a closed orbit that is offset by $\eta\delta$.

where $1/\rho$ is non-zero only in the bends. In competition, the additional speed moves the particle forwards in one turn by

$$\Delta C_{speed} = C\frac{\Delta\beta}{\beta} \tag{4.23}$$

where C is the accelerator circumference, β is the relativistic speed, and

$$\frac{\Delta\beta}{\beta} = \frac{\delta}{\gamma^2} \tag{4.24}$$

relates the fractional change in speed to the nominal γ.

Adding both factors together, the longitudinal displacement z changes according to

$$z_{n+1} = z_n - \eta_s C \cdot \delta_n \tag{4.25}$$

where the slip factor η_s is defined by

$$\eta_s \equiv \frac{1}{\gamma_T^2} - \frac{1}{\gamma^2} \tag{4.26}$$

and the transition gamma γ_T is defined by

$$\frac{1}{\gamma_T^2} \equiv \frac{1}{C}\oint\frac{\eta}{\rho}\,ds \tag{4.27}$$

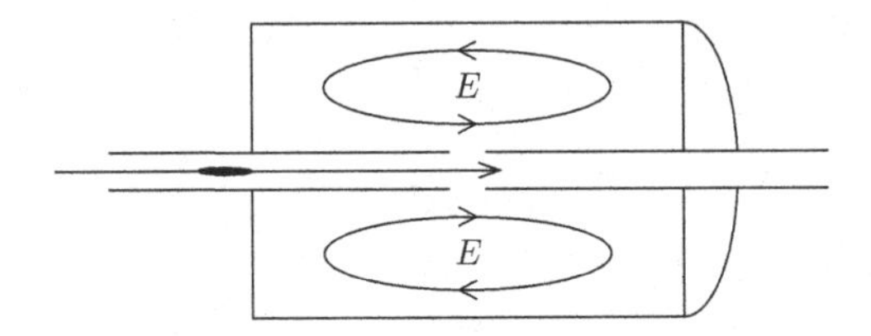

Figure 4.5 A bunch of particles passing through a radio frequency cavity, shown schematically as a section though a simple cylinder with a short accelerating gap. The oscillations of a fundamental electromagnetic mode are synchronised with the turn-by-turn arrival of each bunch.

In the domain of electron accelerators, where $\gamma \to \infty$ and $\eta_s \to 1/\gamma_T^2$, the slip factor is commonly called the (momentum) compaction factor.

Exciting things can happen when hadron accelerators accelerate through the value $\gamma = \gamma_T$. Some hadron accelerators, like RHIC, soften the trauma by employing a transition jump to suddenly lower the value of γ_T at the right time [18], but many others just suffer the consequences, which are usually acceptable. Note that the integral on the right hand side of Equation 4.27 can be arranged to be negative, so that γ_T becomes imaginary, and it is never necessary to cross transition.

Once per turn a particle passes through one or more radio frequency (RF) cavities that contain a resonating electromagnetic field, as shown schematically in Figure 4.5. The particle acquires an additional energy that is proportional to the integrated cavity voltage amplitude V_{RF}, and depends on the phase at the time of arrival according to

$$E_{n+1} = E_n + qV_{RF} \sin\left(2\pi \frac{z_n}{\lambda_{RF}}\right) \tag{4.28}$$

The energy of a nominal particle with $z_n = 0$ on every turn is unchanged if the resonant frequency f_{RF} is an integer multiple of the revolution frequency f_{rev}, so that

$$f_{RF} = \frac{c}{\lambda_{RF}} = hf_{rev} = h\frac{\beta c}{C} \tag{4.29}$$

where βc is the speed of the nominal particle, and the integer h is the harmonic number. Finally, the change in momentum is related to the change in energy by the relativistic result

$$\frac{\Delta p}{p_0} = \frac{1}{\beta^2}\frac{\Delta E}{E_0} \tag{4.30}$$

so that

$$\delta_{n+1} = \delta_n + \left(\frac{qV_{RF}}{\beta^2 E_0}\right) \sin\left(2\pi \frac{z_n}{\lambda_{RF}}\right) \tag{4.31}$$

where E_0 is the nominal energy.

The turn-by-turn evolution of the longitudinal displacement z depends linearly on δ, according to Equation 4.25, while the evolution of δ depends on the sinusoid of z, according to Equation 4.31. Taken together, these two difference equations describe longitudinal motion that is intrinsically nonlinear, thanks to the sinusoid. However, in the limit of small oscillations with $z \ll \lambda_{RF}$ the linearised difference equations of motion become

$$z_{n+1} = z_n - (\eta_s C) \,.\, \delta_n \tag{4.32}$$

$$\delta_{n+1} \approx \delta_n + \left(\frac{qV_{RF}}{\beta^2 E_0}\right)\left(\frac{2\pi}{\lambda_{RF}}\right) . \, z_{n+1} \tag{4.33}$$

These are solved, for small values of the synchrotron tune $Q_s \ll 1$, by

$$z_n = a_z \sin(2\pi Q_s.n + \phi_{s0}) \tag{4.34}$$

$$\delta_n = a_\delta \cos(2\pi Q_s.n + \phi_{s0})$$

where

$$Q_s = \sqrt{\frac{\eta_s}{2\pi} \cdot \frac{C}{\lambda_{RF}} \cdot \frac{qV_{RF}}{\beta^2 E_0}} \tag{4.35}$$

However, the synchrotron tune is not necessarily small – especially in electron accelerators – introducing significantly strong chaos for some oscillation amplitudes. Fortunately, large synchrotron tunes in electron accelerators are accompanied by strong damping, while synchrotron tunes in hadron accelerators (with no damping) tend to be small.

4.4 The Standard Map

The longitudinal motion described by Equations 4.31 and 4.32 is fundamentally the same (within scaling factors) as the standard map

```
until finished {                                    (4.36)
        θ = θ + θ′ . Δt
        θ′ = θ′ − sin(θ) . Δt
}
```

which is also fundamentally the same as the unit length gravity pendulum described in Equation 1.2, when the acceleration due to gravity is scaled to become $g = 1$. Small amplitude motion under the standard map is simply

$$\theta = a \sin(2\pi Q_0.n) \tag{4.37}$$

$$\theta' = a \cos(2\pi Q_0.n) \tag{4.38}$$

where n is the step number, a is the identical amplitude of both θ and θ', and the small amplitude tune Q_0 is given by

$$\cos(2\pi Q_0) = 1 - \frac{\Delta t^2}{2} \tag{4.39}$$

Thus, even small amplitude motion is unstable if $\Delta t > 2$.

Motion under the standard map is fundamentally similar to longitudinal motion in a hadron accelerator – that is, in the absence of significant damping – if the RF cavities are located in a relatively short section of the ring, and are reasonably approximated by a single longitudinal kick per turn. In this case the synchrotron tune Q_s is the standard map tune Q_0, and the behaviour shown in Figure 4.6 is to be expected.

Figure 4.6a – already seen as Figure 1.1 – shows how standard map trajectories with different initial conditions evolve in the limit that $Q_0 \rightarrow 0$, when $\Delta t \rightarrow 0$. Equivalently, it shows the contours of the Hamiltonian function

$$H = \frac{1}{2}p^2 - \cos(q) \tag{4.40}$$

where the co-ordinates p and q are associated with θ' and θ, respectively. The canonical differential equations of motion

$$\frac{dq}{dt} = \frac{\partial H}{\partial p} \tag{4.41}$$

$$\frac{dp}{dt} = -\frac{\partial H}{\partial q} \tag{4.42}$$

reproduce the standard map difference equations of Equation 4.36 in the limit that $\Delta t \rightarrow 0$, since

$$\Delta q = p \,.\, \Delta t \tag{4.43}$$

$$\Delta p = -\sin(q) \,.\, \Delta t \tag{4.44}$$

The canonical equations of motion also guarantee that H is constant in time for any initial conditions, since

$$\frac{dH}{dt} = \frac{\partial H}{\partial p} \,.\, \frac{dp}{dt} + \frac{\partial H}{\partial q} \,.\, \frac{dq}{dt} = 0 \tag{4.45}$$

Thus, motion follows lines of constant H – contours and trajectories are identical.

The velocity of motion along a contour is

$$\vec{v} = \frac{dp}{dt}\widehat{p} + \frac{dq}{dt}\widehat{q} \tag{4.46}$$

$$= -\frac{\partial H}{\partial q}\widehat{p} + \frac{\partial H}{\partial p}\widehat{q} \tag{4.47}$$

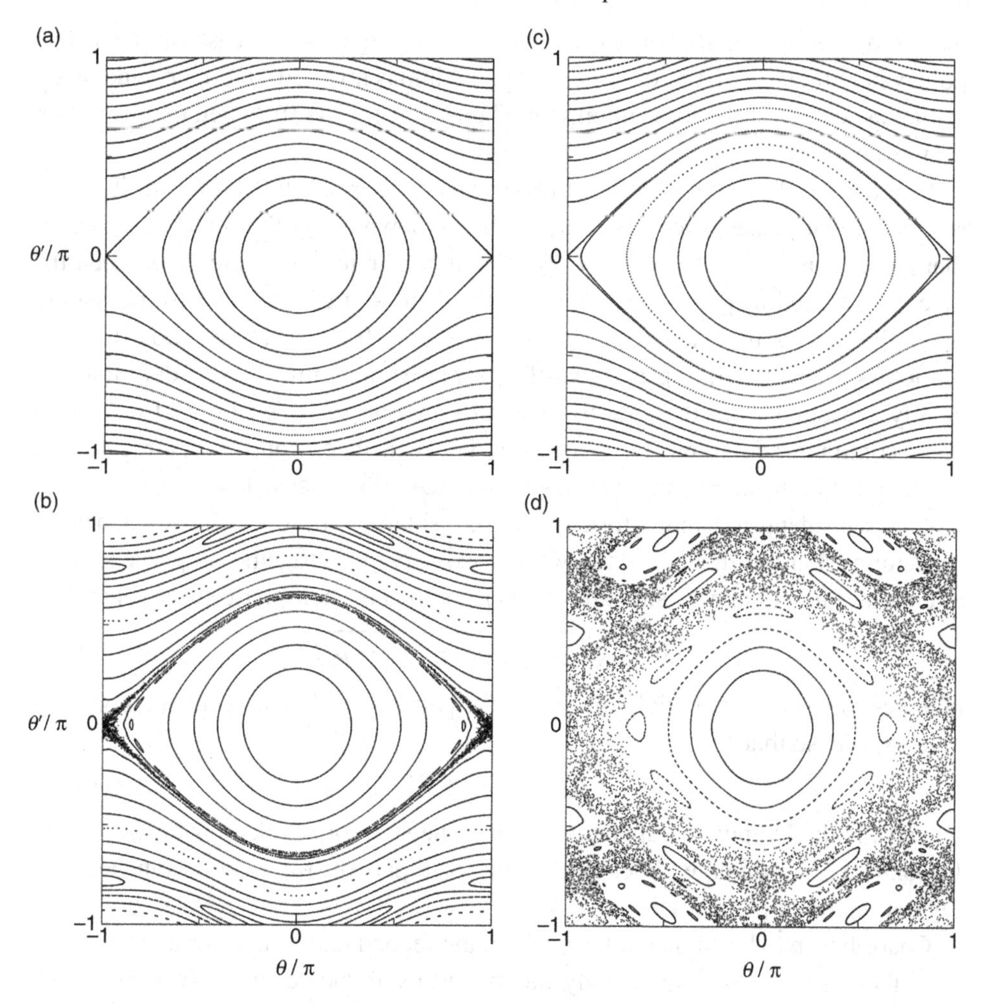

Figure 4.6 Regular and chaotic behaviour under the standard map with increasing tunes, in each case for many trajectories launched with different initial conditions. Contours of the Hamiltonian $H = 1/2\theta'^2 - \cos(\theta)$ are shown in (a), in the canonical limit of the gravity pendulum when $Q_0 \to 0$. Regular motion is visible everywhere in (b), when $Q_0 = 0.06$. Chaos starts to appear when $Q_0 = 0.12$ in (c), near what was the pendulum separatrix, contained by regular Kolmogorov–Arnold–Moser surfaces. Large areas of chaotic sea contain many resonance islands in (d), when $Q_0 = 0.18$.

where $\widehat{p}$ and $\widehat{q}$ are unit vectors, while the slope of the Hamiltonian is

$$\vec{\nabla} H = \frac{\partial H}{\partial p}\widehat{p} + \frac{\partial H}{\partial q}\widehat{q} \tag{4.48}$$

showing that these two vectors are perpendicular, but have the same length. Thus, the speed of motion along a Hamiltonian contour is equal to the steepness of the function, visually represented by the closeness of the contours. For example, all motion of the equivalent pendulum represented in Figure 4.6a ceases at the saddle point where $(\theta, \theta') = (\pm\pi, 0)$ (when the pendulum is inverted, stationary, and metastable) as well as at the local minimum where $(\theta, \theta') = (0, 0)$ (when the pendulum is at rest and stable).

Inverted pendula are sensitive to the smallest perturbations, and so it is not surprising that phase space first starts to break down along the separatrix contour that passes through $(\theta, \theta') = (\pm\pi, 0)$. This is illustrated in Figure 4.6c, when the tune is increased to $Q_0 = 0.12$. Chaos spreads to a large fraction of phase space when the tune is increased even further, to $Q_0 = 0.18$, in Figure 4.6d. Chaotic motion is visually striking as seas of apparently randomly placed dots that are clearly separated from the regions of regular motion in which lines of dots form quasi-continuous contours. The boundaries between chaotic and regular regions are properly called Kolmogorov–Arnold–Moser (KAM) surfaces [12].

Chaos is identified more formally by considering the evolution of a pair of trajectories A and B that are initially launched very close together in phase space – as close as numerical precision allows, on a computer. Their phase space separation

$$\Delta_{AB} = \sqrt{(q_A - q_B)^2 + (p_A - p_B)^2} \tag{4.49}$$

diverges roughly exponentially in time if the pair is launched in a chaotic region of phase space, so that

$$\Delta_{AB}(n) \approx k\, e^{\lambda n} \tag{4.50}$$

where λ is the Lyapunov exponent, and n is the step (or accelerator turn) number. In contrast, Δ_{AB} diverges linearly in time for a pair of trajectories launched in a regular region of phase space.

Chaos began to be studied intensively in the second half of the twentieth century, even though the topic was already mentioned by Poincaré in the late nineteenth century [44]. Nonetheless, practically viable research into nonlinear difference systems, and chaos, had to wait for the advent of computers. Fortunately, computers and our understanding of chaos arrived in time to address the practical challenges of the nonlinear dynamics of accelerators.

Exercises

4.1 Suppose that a single dipole placed centrally between thin D and F quadrupoles in a FODO half-cell has a bend angle θ, but occupies only a fraction of the available space L.

a) How do the maximum and minimum matched dispersion functions compare to a FODO cell with 100% dipole occupancy?

b) What happens if there are N evenly distributed dipoles in each half-cell, with the same total bend angle θ?

4.2 Consider the close connection between η and β in well-matched cells.

a) How does the value of $\eta/\sqrt{\beta}$ evolve through a FODO half-cell, for a typical value of phase advance per cell $\Delta\phi$, say 60 or 90 degrees?
b) How far does $\eta/\sqrt{\beta}$ deviate from a constant value?

4.3 Suppose that the following one-turn matrix M transforms motion from $s = 0$ to $s = C$ around the circumference of an accelerator:

$$\begin{pmatrix} -1.05746 & -3.59421 & 0.00000 & 0.00000 & 0.00000 & 35.44680 \\ 0.00189 & -0.93923 & 0.00000 & 0.00000 & 0.00000 & -0.82369 \\ 0.00000 & 0.00000 & 1.72622 & -72.45113 & 0.00000 & 0.00000 \\ 0.00000 & 0.00000 & 0.05149 & -1.58161 & 0.00000 & 0.00000 \\ -0.80399 & -36.25338 & 0.00000 & 0.00000 & 1.00000 & -50.03916 \\ 0.00000 & 0.00000 & 0.00000 & 0.00000 & 0.00000 & 1.00000 \end{pmatrix}$$

a) What are the fractional parts of the horizontal and vertical tunes?
b) What are the horizontal Twiss functions (β, α, γ) at $s = 0$?
c) What is the dispersion function and its slope, η and η', at $s = 0$?

4.4 (See also Exercises 5.8 and 11.8.) The circumference of RHIC is 3.834 km, and the harmonic number of the high frequency RF system is 7×360, with transition $\gamma_T = 22.89$. Assume that the total RF voltage is 6 MV.

a) What is the slip factor for gold ions circulating with $\gamma = 100$?
b) If the gold ions are fully stripped, what is the synchrotron tune?

4.5 Derive Equation 4.39.

4.6 (See also Exercise 9.7.) Investigate motion under the standard map by writing code that permits you to adjust the tune Q_0, and allows you to follow and record the motion of two trajectories launched very close together, at any initial location in phase space.

a) Confirm that the two trajectories diverge linearly in time when the motion is regular, and exponentially when the motion is chaotic.
b) Plot the Lyapunov exponent λ as a function of Q_0 for a set of several initial phase space locations of particular interest.
c) Is it possible to automate the identification of the motion as regular or chaotic, for particular initial phase space locations that are input to an algorithm?
d) Are there any values of Q_0 where regular motion completely disappears, everywhere?

5

Action and Emittance – One Particle or Many?

An alternative title for this chapter is *One particle, many, or none?* The Twiss functions β, α and γ were first introduced in Equation 3.16 as a convenient way to parameterise M, the one-turn matrix at an arbitrary reference point. As such, one-turn Twiss parameters are purely a property of the optical lattice, with no direct need to invoke even a single particle, let alone many. Later, one-turn Twiss parameters at two locations were used to parameterise the matrix M_{21} that describes how single particles propagate from s_1 to s_2 (Equation 3.31). Then the tables were turned, instead using the matrix M_{21} to describe how the one-turn Twiss parameters themselves propagate from s_1 to s_2, (Equation 3.34).

Another interpretation sees Twiss parameters as properties of the beam, and not of the optics. Beam Twiss parameters describe the distribution properties, such as size, of bunches of very many particles. From this perspective, Equation 3.34 describes the propagation of the distribution from s_1 to s_2 – a perspective that is often natural in single-pass systems such as linacs and transfer lines, in which it does not make sense to apply periodic boundary conditions. Beam Twiss parameters are also useful in rings, describing the distribution of a newly injected bunch during the relatively short time in which initial transients are dying down, before arriving at a stable distribution.

Asking whether Twiss parameters describe optics or beam is like asking whether a photon is a wave or a particle: the answer is ‘both’ (or ‘yes’).

5.1 Transverse Action-Angle Co-ordinates

Written in terms of action-angle co-ordinates J_x and ϕ_x, the (linearised) horizontal displacement and angle of a particle on turn n in a ring are

$$\begin{aligned} x_n &= \sqrt{2J_x\beta_x}\ \sin(\phi_x) \\ x'_n &= \sqrt{2J_x/\beta_x}\ [\cos(\phi_x) - \alpha_x \sin(\phi_x)] \end{aligned} \tag{5.1}$$

where action J_x is a constant and the phase is

$$\phi_{x,n} = 2\pi Q_x n + \phi_0 \tag{5.2}$$

The Courant–Snyder invariant $2J_x$ that labels a test particle is found from a single pair of co-ordinates (x, x') at *any* reference point around the ring, on *any* turn n, by inverting Equation 5.1 and using the Twiss identity

$$\beta\gamma = 1 + \alpha^2 \tag{5.3}$$

that guarantees the one-turn matrix M has unit determinant. This gives an equation

$$2J_x = \beta_x x'^2 + 2\alpha_x xx' + \gamma_x x^2 \tag{5.4}$$

that also describes the ellipse illustrated in Figure 5.1.

The set of (x, x') values that forms the ellipse can represent the turn-by-turn motion of a single particle over many turns, following the linear motion described by Equations 5.1 and 5.2. Alternatively, the body or the edge of the ellipse can be taken to represent a set of many particles that are 'typical' of all those in a single bunch, in a snapshot taken at a reference point in the ring on a single turn. Either way, the area of the ellipse is

$$A_x = 2\pi J_x \tag{5.5}$$

everywhere around the ring, even though the ellipse stretches and distorts as the reference point moves.

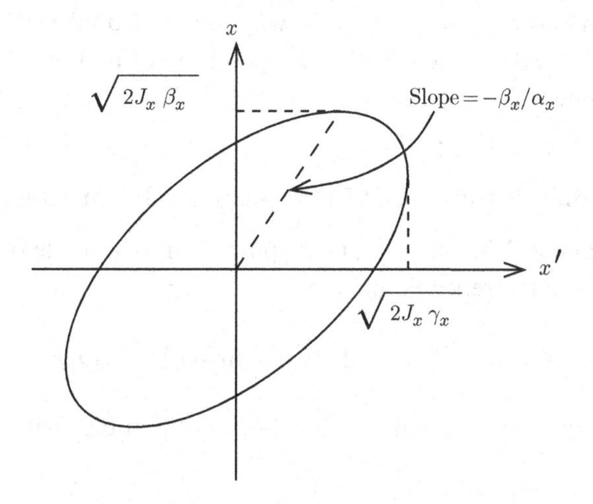

Figure 5.1 The phase space ellipse described by the Twiss parameters (β, α, γ) and the action J_x, at a reference point in a ring.

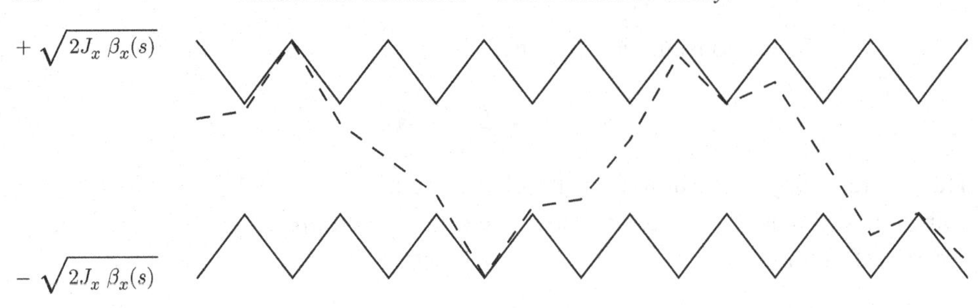

Figure 5.2 The horizontal displacement (dashed line) of a single on-momentum particle oscillating within a betatron envelope boundary of $\pm\sqrt{2J_x\beta_x(s)}$. The envelope (solid line) shown here is from a continuous channel of FODO cells with a phase advance of 80 degrees per cell.

Turn-by-turn motion is even simpler using the normalised phase space co-ordinates introduced in Section 3.2, since then

$$\begin{pmatrix} \tilde{x} \\ \tilde{x}' \end{pmatrix}_n = \sqrt{2J_x} \begin{pmatrix} \sin(\phi_{x,n}) \\ \cos(\phi_{x,n}) \end{pmatrix} \tag{5.6}$$

and the ellipse becomes a circle, with the same area A_x.

Propagation around a fraction of a turn is written

$$\begin{aligned} x(s) &= \sqrt{2J_x\beta_x(s)}\;\; \sin(\phi_x(s)) \\ x'(s) &= \sqrt{2J_x/\beta_x(s)}\;\; [\cos(\phi_x(s)) - \alpha_x(s)\sin(\phi_x(s))] \end{aligned} \tag{5.7}$$

in close similarity to Equation 5.1, but where now the modulations of β_x and α_x with s, and the monotonic increase of ϕ_x with s, are also taken into account. The confinement of a single particle within an envelope of amplitude $\pm\sqrt{2J_x\beta_x(s)}$ is illustrated in Figure 5.2.

5.2 Unnormalised Emittances and Beam Sizes

The root mean square (RMS) horizontal displacement of a single particle, observed over many turns at a reference point, is

$$\sigma_{x,1} \equiv \langle x^2 \rangle^{1/2} = 2\sqrt{J_x\beta_x}\,\langle \sin^2(\phi_x)\rangle = \sqrt{J_x\beta_x} \tag{5.8}$$

Thus, the mean square horizontal size of a bunch of particles with an action distribution $\rho(J_x)$ is

$$\sigma_x^2 = \frac{1}{N}\int_0^\infty \sigma_{x,1}^2\, \rho(J_x)\, dJ_x \tag{5.9}$$

where the population of the bunch

$$N = \int_0^\infty \rho(J_x)\, dJ_x \tag{5.10}$$

is typically as large as 10^{11} protons, or 10^9 gold ions. It is natural to define the unnormalised RMS emittance $\epsilon_{x,u}$ as the average action

$$\epsilon_{x,u} \equiv \langle J_x \rangle \tag{5.11}$$

because then the mean square horizontal beam size is simply

$$\sigma_x^2 = \beta_x\, \epsilon_{x,u} \tag{5.12}$$

with a similar expression in the vertical.

Typical horizontal beam sizes in many rings – although by no means all – are of order 1 mm, while β-functions are often in the range 10 to 100 m, in both planes. Unnormalised horizontal RMS emittances are therefore often in the range of 10 to 100 nm. Vertical emittances are usually commensurate with horizontal emittances (the beams are ‘round’) in hadron rings, but they are often orders of magnitude smaller (the beams are ‘flat’) in electron storage rings.

For completeness, all three horizontal second-order beam distribution moments are

$$\begin{pmatrix} \langle x^2 \rangle \\ \langle xx' \rangle \\ \langle x'^2 \rangle \end{pmatrix} = \epsilon_{u,x} \begin{pmatrix} \beta_x \\ -\alpha_x \\ \gamma_x \end{pmatrix} \tag{5.13}$$

with a similar expression in the vertical when x is replaced by y.

Particles perform simultaneous oscillations in all three dimensions, and coupling can occur between them. Fortunately it is often possible to neglect these couplings, especially after the three tunes are properly set, and horizontal-vertical coupling is corrected. In the ideal case when horizontal and vertical oscillations are perfectly decoupled, then the cross term moments

$$\langle xy \rangle = \langle xy' \rangle = \langle x'y \rangle = \langle x'y' \rangle = 0 \tag{5.14}$$

are all zero. Although this is never perfectly true everywhere around a ring, it is often a good approximation.

Assuming there is dispersion η in the horizontal plane but none in the vertical, then the total horizontal and vertical displacements are

$$\begin{aligned} x_{TOT,n} &= \sqrt{2J_x\beta_x}\ \sin(\phi_{x,n}) \ + \ \eta\, \delta_n \\ y_n &= \sqrt{2J_y\beta_y}\ \sin(\phi_{y,n}) \end{aligned} \tag{5.15}$$

If the relative momentum error on turn n is

$$\delta_n = \frac{\Delta p}{p_0} = a_\delta \sin(\phi_{s,n}) \tag{5.16}$$

and the horizontal betatron and synchrotron phases ϕ_x and ϕ_s are uncorrelated, then the total horizontal RMS beam size is

$$\sigma^2_{x,TOT} = \beta_x \epsilon_{x,u} + \eta \left(\frac{\sigma_p}{p_0}\right)^2 \tag{5.17}$$

where σ_p is the RMS momentum spread. The simplicity of this sum-of-squares addition is one advantage of using a definition of emittance that does not assume a particular beam distribution shape (for example Gaussian), and which does not rely on 'typical' ellipses that contain a certain fraction of the beam.

5.3 Tune Spread and Filamentation

Often the emittance – the intrinsic beam size – must be kept as small as possible, for example during beam transfer from one ring to another. This is illustrated by the simple simulation results presented in Figure 5.3 for the exaggerated case of a small emittance beam entering a ring with a large displacement error of $\Delta x = 2$ mm at

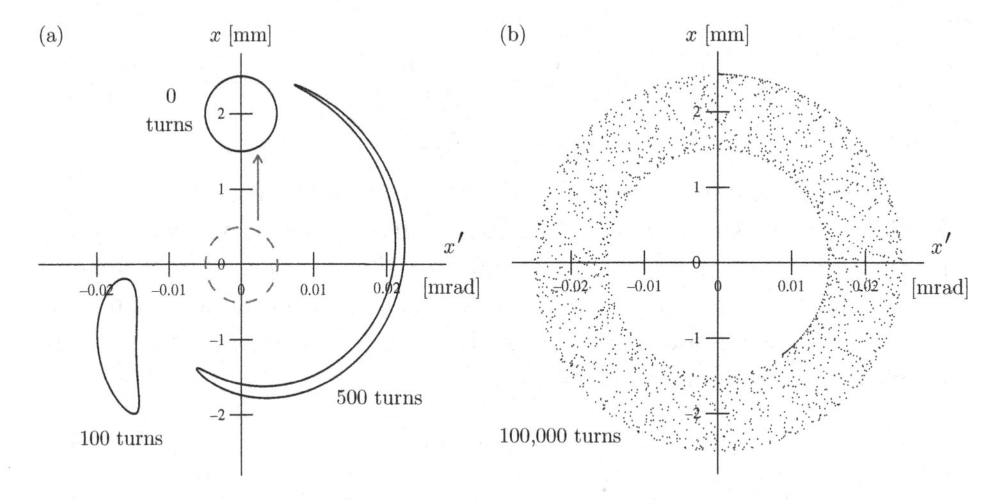

Figure 5.3 Phase space filamentation of a badly injected beam, at a reference point with $\beta_x = 100$ m and $\alpha = 0$. The black circle in (a) represents the incoming beam, with 1,800 particles evenly distributed around its azimuth, after being displaced from the origin by an error displacement of 2 mm. The tear drop and the crescent in (a) show how the distribution filaments after 100 and 500 turns, due to a total tune spread of about $\Delta Q = 0.001$ between the smallest and largest actions. No significant azimuthal structure is left after 100,000 turns in (b), when filamentation is complete.

a reference point with $\beta = 100$ m and $\alpha = 0$. The incoming beam contains 1,800 particles that are intended to be evenly distributed in phase space on the dashed circle around the origin of Figure 5.3a. Ideally all particles would have the same action, and the emittance would be $\epsilon_{u,0} = \langle J \rangle = 1.25$ nm.

However, the 2 mm injection displacement error moves the dashed circle vertically in phase space, to the solid circle that is an initial scatter plot of all $N = 1,800$ simulated particles. The error is large enough, compared to the incoming RMS beam size of 0.35 mm, that the initial action-angle distribution is approximately a double δ-function:

$$\rho_{initial}(J, \phi) \approx N\,\delta(J - J_{inj})\,\delta(\phi - \phi_{inj}) \tag{5.18}$$

where $J_{inj} = 20$ nm and $\phi_{inj} = \pi/2$. If all the injected particles have exactly the same tune, then the circle of particles maintains its shape as it rotates coherently around the phase space origin. However, this is not realistic. In practice (and as discussed in more detail, in later chapters) the tune varies linearly with action, so that

$$Q(J) = Q(0) + \frac{dQ}{dJ}.J \tag{5.19}$$

where dQ/dJ is approximately constant. Tune spread causes azimuthal shear: individual particles still rotate (turn-by-turn) on circles around the phase space origin, but larger and smaller amplitude particles rotate at somewhat different speeds.

The total tune spread in the simulation is about 0.001, from the smallest action to the largest. After 100 turns the circular distribution has evolved to a teardrop, and after 500 turns the scatter plot has filamented to a thin crescent with an angular spread of about π radians. However, the area enclosed by the apparently continuous chain of particles remains constant during the shearing! This is a consequence of Liouville's theorem, interpreted by Mackay [11] as:

In the local region of a particle, the particle density in phase space is constant, provided that the particles move in a general field consisting of magnetic fields and of fields whose forces are independent of velocity.

The formal certainty of area preservation is rather moot after 100,000 turns, when filamentation is over and the transients are dead, as shown in Figure 5.3b. For all practical purposes the injected particles are spread smoothly over a final phase space area that is much larger than the initial area. In summary, there is no structure in angle ϕ in the final distribution

$$\rho_{final}(J, \phi) = \rho_{final}(J) \approx N\,\delta(J - J_{inj}) \tag{5.20}$$

but the average action is increased from 1.25 nm to 21.25 nm. (See Section 5.3 for a more exact discussion of injection errors.)

The (practical) area containing the beam increases by more than a factor of 20 from injection until equilibrium, in this example, with an ill-defined value while the distribution is still evolving. In stark contrast, the average action remains constant after the increase that occurs instantaneously upon injection. Defining the unnormalised emittance through the average action is natural for rings, in part because the details of the evolution of beam distribution transients can then simply be ignored.

5.4 Linac (Phase Space Area) Emittances

Transients cannot be ignored in linacs or in other single-pass systems, because there is no second turn! One-turn Twiss functions defined by the lattice optics cannot be invoked. Instead, it is natural to use beam Twiss functions, and it is common to derive emittances from phase space areas.

When the horizontal second-order beam distribution moments $\langle x^2 \rangle$, $\langle xx' \rangle$ and $\langle x'^2 \rangle$ are available – either by real world measurement or from a scatter-plot simulation – then the beam Twiss parameters are derived by inverting Equation 5.13 to give

$$\begin{pmatrix} \beta_x \\ -\alpha_x \\ \gamma_x \end{pmatrix} = \frac{1}{\epsilon_{x,A}} \begin{pmatrix} \langle x^2 \rangle \\ \langle xx' \rangle \\ \langle x'^2 \rangle \end{pmatrix} \tag{5.21}$$

The Twiss identity (Equation 5.3) is guaranteed by defining the area emittance to be

$$\epsilon_{x,A} \equiv \sqrt{\langle x^2 \rangle \langle x'^2 \rangle - \langle xx' \rangle^2} \tag{5.22}$$

so the ellipse that represents the beam distribution

$$\epsilon_{x,A} = \beta_x x'^2 + 2\alpha_x xx' + \gamma_x x^2 \tag{5.23}$$

has maximum extents

$$\pm \sqrt{\epsilon_{x,A}\beta_x} = \pm \langle x^2 \rangle^{1/2} \tag{5.24}$$

on the x-axis, and

$$\pm \sqrt{\epsilon_{x,A}\gamma_x} = \pm \langle x'^2 \rangle^{1/2} \tag{5.25}$$

on the x'-axis, with an area

$$A_x = \pi \, \epsilon_{x,A} \tag{5.26}$$

Associating the emittance so closely with the area of a representative ellipse has its difficulties, however.

Wangler [57] illustrates one of the difficulties with the simple example of a beam distribution

$$x' = C\,x^m \tag{5.27}$$

where C is a constant and m is a positive integer. The phase space area of the beam distribution is zero, since Equation 5.27 is a one-dimensional line (straight if $m = 1$). In general, however, the area emittance of the representative ellipse

$$\epsilon_{x,A} = C\,\sqrt{\langle x^2\rangle\langle x^{2m}\rangle - \langle x^{m+1}\rangle^2} \tag{5.28}$$

is non-zero unless $m = 1$. Wangler concludes:

The [area] emittance depends not only on the true area occupied by the beam in phase space, but also on the distortions produced by nonlinear forces.

Injection Displacement and Momentum Errors

If the injected beam is horizontally displaced by Δx from the closed orbit of an accelerator, every individual particle is shifted by $\Delta x/\sqrt{\beta}$ in normalised phase phase, as illustrated by Figure 5.4. Thus, the total action of a particle that initially had action-angle co-ordinates (J, ϕ) becomes

$$J_T = J + \Delta x\sqrt{\frac{2J}{\beta}}\cos(\phi) + \frac{\Delta x^2}{\beta} \tag{5.29}$$

Averaging over all particles, and noting that $\langle\cos(\phi)\rangle = 0$ after filamentation is complete, the new emittance increases to

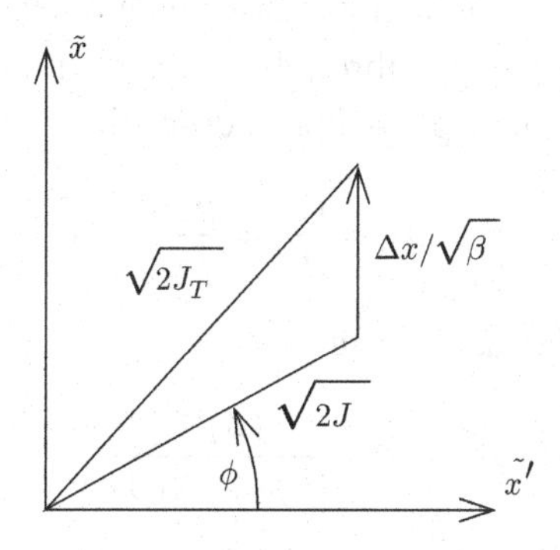

Figure 5.4 Normalised phase space geometry of a particle that is injected with action-angle co-ordinates (J, ϕ), including a pure displacement error of Δx that increases its action to J_T.

$$\epsilon_{u,T} = \langle J_T \rangle = \epsilon_{u,0} + \frac{\Delta x^2}{2\beta} \tag{5.30}$$

or, equivalently

$$\epsilon_{u,T} = \epsilon_{u,0} \left(1 + \frac{\Delta x^2}{2\sigma_0^2} \right) \tag{5.31}$$

showing that the emittance growth varies like the square of the displacement error scaled with σ_0, the incoming RMS beam size. In the exaggerated case discussed in Figure 5.3, the emittance increases from 1.25 nm to 21.25 nm.

Even when the injected beam is perfectly aligned, the average momentum may be wrong. Since the total horizontal displacement of a particle is

$$x_{TOT} = \eta\,\delta + x \tag{5.32}$$

then a momentum error of δ_{error} is equivalent to a displacement error of

$$\Delta x = -\eta\,\delta_{error} \tag{5.33}$$

and the new horizontal emittance increases to

$$\epsilon_{u,T} = \epsilon_{u,0} + \frac{\eta^2}{2\beta}\,\delta_{error}^2 \tag{5.34}$$

where angular errors in normalised phase space are assumed to be unimportant (for example if $\eta' = 0$). It is clearly a good idea to design dispersion-free injection straights with $\eta = \eta' = 0$, so that injection momentum errors do not contribute to the horizontal emittance.

5.5 Normalised Emittance and Adiabatic Damping

Suppose that each particle in a bunch increases its momentum along the accelerator axis by $m_0 c\,\Delta(\beta\gamma)$ during passage through an accelerating RF cavity, as illustrated in Figure 5.5. Since the angular phase space co-ordinate of the particle is

$$x' = \frac{dx}{ds} = \frac{p_\perp}{p_\parallel} \tag{5.35}$$

then all angles shrink during acceleration

$$x' \rightarrow x' \frac{(\beta\gamma)}{(\beta\gamma) + \Delta(\beta\gamma)} \tag{5.36}$$

and the action of each particle is modified according to Equation 5.4.

The modified action is usually also smaller than the original, but it may be larger, depending on the Twiss values (β, α, γ) at the RF cavity and on the phase ϕ of the

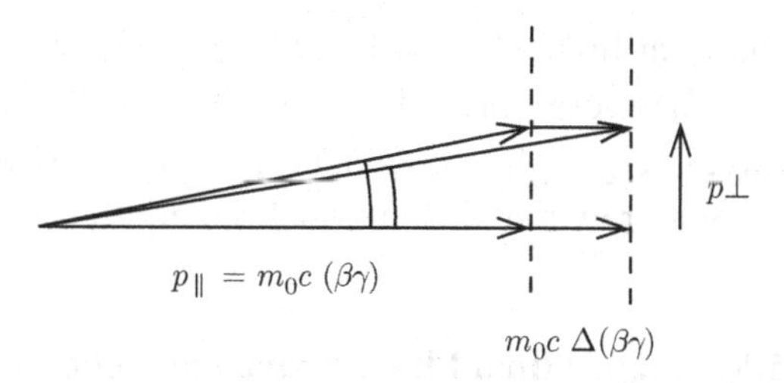

Figure 5.5 Shrinkage of the angular phase space co-ordinate $x' = dx/ds$ when the momentum parallel to the accelerator axis increases by $m_0c\,\Delta(\beta\gamma)$ during passage through an accelerating RF cavity.

test particle. Nonetheless, averaging over the phases of many particles all with the same initial action J

$$\langle J_{new}\rangle = \frac{J}{1 + \Delta(\beta\gamma)/(\beta\gamma)} \tag{5.37}$$

shows that their average action always decreases. Averaging over the action of all particles, it follows that the unnormalised emittance ϵ_u shrinks adiabatically during acceleration according to

$$\epsilon_u = \frac{\epsilon_n}{\beta\gamma} \tag{5.38}$$

where the normalised emittance ϵ_n is a constant in the idealised situation when there are no sources of noise that increase the emittance.

The equation for the RMS beam size due to betatron oscillations

$$\sigma = \sqrt{\beta\epsilon_u} = \sqrt{\frac{\beta\epsilon_n}{(\beta\gamma)}} \tag{5.39}$$

introduces the potential for confusion: β stands for the Twiss function in the numerator, but for the Lorentz quantity in the denominator. Confusion also exists in the literature of hadron accelerators, unfortunately, with other definitions of transverse emittance. A factor of 4π or 6π is often seen in the denominator of Equation 5.39, for example, after defining emittance as the area in phase space that contains 95% of the particles in a bunch. These other definitions implicitly assume a particular beam distribution shape, usually Gaussian. Such an assumption is not necessary when the RMS emittance is defined through average action.

Electrons circulating an accelerator emit a large number of synchrotron radiation photons, as discussed in Chapter 11. A dynamic equilibrium is reached between the damping and excitation effects caused by the synchrotron radiation, establishing an unnormalised RMS emittance ϵ_u that is a function of both the energy and the lattice optics of the accelerator. The characteristic time it takes to reach equilibrium is quite short, especially at higher energies when copious photons are produced. Adiabatic shrinkage is therefore only relevant in electron accelerators on shorter timescales, and at lower energies. In consequence, the literature of circular electron (or positron) accelerators rarely mentions normalised emittance ϵ_n.

5.6 Longitudinal Phase Space Parameters

The canonical co-ordinates often used to study longitudinal motion in an RF bucket are (ϕ_{RF}, W), where ϕ_{RF} is the RF phase of a test particle, and

$$W = \frac{E - E_s}{\omega_{RF}} \tag{5.40}$$

is the total energy offset of the test particle relative to the nominal synchronous particle, scaled by ω_{RF}, the angular frequency of the RF system. If most of the particles are well inside the RF bucket separatrix, then the RMS bunch area is conveniently defined as

$$S_{RMS} = \pi\, \sigma_W \sigma_\phi \quad [\mathrm{eV\,s}] \tag{5.41}$$

where σ_W and σ_ϕ are the standard deviations of the distribution.

The 95% bunch area that is often seen in the literature is related to the RMS area of a Gaussian bunch through $S_{95} = 6\,S_{RMS}$. The factor of π that inevitably enters when phase space area is discussed (in place of average action) is usually *explicitly* included when numerical values are quoted, so it is common to see '$S_{RMS} = 0.1$ [eV s]' and rare to see '$S_{RMS} = 0.032\,\pi$ [eV s]', expressions which are approximately identical.

RMS bunch length σ_s and the relative momentum spread σ_p are often the practical parameters of choice, rather than σ_W and σ_ϕ. It is readily shown that

$$\sigma_W \sigma_\phi = \frac{m_0 c^2}{c}\,(\beta\gamma)\,\sigma_s \sigma_p \tag{5.42}$$

where m_0 is the rest mass of the test particle – or the rest mass per nucleon, in the case of ions [42].

The normalised longitudinal emittance ϵ_s and the longitudinal beta function β_s are defined in direct analogy to their transverse counterparts

$$\sigma_s = \sqrt{\frac{\beta_s \epsilon_s}{(\beta\gamma)}} \tag{5.43}$$

and

$$\sigma_p = \sqrt{\frac{\epsilon_s}{\beta_s(\beta\gamma)}} \tag{5.44}$$

where $\beta\gamma$ is the usual Lorentz factor. These equations may also be written as

$$\epsilon_s \,[\mathrm{m}] = \sigma_s \sigma_p \,(\beta\gamma) = \frac{S}{\pi} \frac{c}{m_0 c^2} \tag{5.45}$$

and

$$\beta_s \,[\mathrm{m}] = \frac{\sigma_s}{\sigma_p} \tag{5.46}$$

The longitudinal beta function is a property of the optics that does not vary with azimuth around a ring, since σ_s and σ_p themselves do not vary. Remarkably, it is given by the simple expression

$$\beta_s = \frac{C}{2\pi} \frac{|\eta_s|}{Q_s} \tag{5.47}$$

where C is the circumference and Q_s is the synchrotron tune [11].

The slip factor η_s is also a fixed property of the transverse optics, especially when the relativistic γ is very large (usually the case with electrons), or at least if γ has a constant value for hadrons. In this case the synchrotron tune and the longitudinal emittance, Q_s and ϵ_s, can be used as the two natural independent variables that, together, summarise the distribution of beam in longitudinal phase under normal conditions.

Clearly, something dramatic happens in the longitudinal phase space of a hadron accelerator when transition is crossed during acceleration, since in that case $\eta_s \to 0$ and $\beta_s \to 0$ [58].

Exercises

5.1 Derive Equation 5.4.

5.2 If the horizontal projection of the beam distribution is Gaussian at a point with zero dispersion $\rho(x) \sim e^{-x^2/2\sigma^2}$, what is the shape of $\rho(J)$, the action distribution?

5.3 Derive Equation 5.29. A beam is injected with an unnormalised emittance of $\epsilon_{u,0}$ at a location with arbitrary β and α, with no displacement error but with an angular error of $\Delta x'$. What is the emittance increase?

5.4 Suppose that incoming beam is injected with perfect alignment, and with the correct momentum. However, the periodic Twiss β-function has an error of $\Delta\beta$. What is the emittance increase? If necessary, assume for simplicity that $\alpha = 0$ in design and in reality.

5.5 (See also Exercise 2.5.) You inherit a set of identical quadrupoles 0.5 m long from a defunct accelerator. The quadrupoles have a 70 mm bore radius, and a maximum pole tip field of 1.1 T.

a) What is the minimum focal length (measured from the longitudinal centre) for a $p = 15$ GeV/c proton beam?
b) You build a FODO line to transport this beam with a phase advance of 60 degrees per cell in each plane. What is the closest apart that you can place consecutive F and D quadrupoles?
c) What are the minimum and maximum β-functions?
d) What is the largest RMS emittance that can be accepted, if $\pm 4\sigma$ of the beam must be transmitted without scraping?

5.6 Prove Equation 5.37.

5.7 Derive Equation 5.47.

5.8 (See also Exercises 4.4 and 11.8.) The circumference of RHIC is 3.834 km, and the harmonic number of the high frequency RF system is 7×360, with transition $\gamma_T = 22.89$. Assume that protons circulate with $\gamma = 250$, and that the total RF voltage is 6 MV.

a) What is the synchrotron tune?
b) What is the longitudinal beta function β_s?
c) If the RMS bunch length is $\sigma_s = 0.1$ m, what is the RMS relative momentum spread?
d) Clearly, σ_s must be significantly less than the RF wavelength. What (approximately) is the maximum longitudinal emittance (in meters) that can be safely contained within an RF bucket?
e) What is the corresponding 95% bunch area S_{95} (in eV-s)?

5.9 Multiple Coulomb scattering occurs as a beam passes through a thin beam instrumentation flag, adding an RMS angular divergence of

$$\bar{\theta} = \sqrt{\langle\theta^2\rangle} \approx z\left(\frac{20 \text{ MeV/c}}{p\beta}\right)\sqrt{\frac{x}{L_{\text{rad}}}}\left(1 + \frac{1}{9}\log_{10}\frac{x}{L_{\text{rad}}}\right) \tag{5.48}$$

where z is the beam particle charge, x is the flag thickness (traditionally measured in [g cm^{-2}]), and L_{rad} is the radiation length of the flag material. This increase in angular divergence enlarges the emittance.

A gold beam with a total energy of 10 GeV/nucleon passes through a 1 mm thick Al_2O_3 flag, at a location in a transfer line where $\beta = 6$ m and $\alpha = 0$. Assume that $L_{\text{rad}} = 24$ g cm^{-2} and $\rho_{Al_2O_3} = 3.7$ g cm^{-3}.

a) Evaluate $\bar{\theta}$.
b) If the normalised RMS emittance is $\epsilon_n = 2$ μm just upstream of the flag, what is the downstream emittance?

6

Magnets

Linear magnets – dipoles and quadrupoles – were introduced in Chapter 2, followed in Chapter 3 by the development of a formalism that describes linear motion through them, in the ideal world of optics design. The discussion is generalised here, to include an extended set of normal and skew multipole magnets, and their potential field errors. This sets the stage for a description of the consequences and correction of some real-world errors, in Chapter 8.

6.1 Normal and Skew Multipole Magnets

Maxwell's equations 2.8 and 2.9 are solved in a multipole expansion by

$$B_y + iB_x = \sum_{n=0}^{\infty} c_n (x+iy)^n \tag{6.1}$$

under the assumption that there are no longitudinal magnetic fields ($B_s = 0$), that the field does not vary with time ($d\vec{B}/dt = 0$), and that there are no current source terms in the region of interest around the beam ($\vec{J} = 0$). The multipole coefficient c_n is a complex number, conventionally written as

$$c_n = b_n + ia_n \tag{6.2}$$

where b_n and a_n are both real, and $i = \sqrt{-1}$. (Many European authors use an alternative convention, preferring to write the right hand side of Equation 6.1 as $\sum_{n=1}^{\infty} c_n(x+iy)^{n-1}$.)

A *normal* magnet with $a_n = 0$ for all n has a purely vertical field on the horizontal mid-plane, since when $y = 0$

$$B_y + iB_x = \sum_{n=0}^{\infty} b_n x^n + i\,0 \tag{6.3}$$

and the field is a polynomial expansion of the horizontal displacement. Only b_0 is non-zero for the normal dipole sketched in Figure 2.5, so that

$$B_y + iB_x = b_0 + i\,0 \tag{6.4}$$

and only b_1 is non-zero for the normal quadrupole of Figure 2.6, when

$$B_y + iB_x = b_1(x + i\,y) \tag{6.5}$$

Combined function magnets have more than one non-zero b_n – usually just b_0 and b_1, but sometimes including b_2 – so that the magnetic field is a sum of multipoles.

A *skew* magnet has all b_n zero. If only one value of a_n is non-zero, then the skew multipole field is

$$B_y + iB_x = i\,a_n(x + iy)^n \tag{6.6}$$

so that the field on the horizontal mid-plane is purely horizontal. The opposite is not true: on the vertical mid-plane, with $x = 0$, the field is either purely horizontal or purely vertical, depending on whether n is odd or even.

In many cases a magnet is designed to be a pure normal (or skew) multipole, with b_n (or a_n) non-zero for only a single value of n. The field of a pure normal multipole, using the polar co-ordinate system (r, θ) shown in Figure 6.1, is

$$\begin{aligned} B_y + iB_x &= b_n\, r^n\, e^{in\theta} \\ &= b_n\, r^n(\cos(n\theta) + i\sin(n\theta)) \end{aligned} \tag{6.7}$$

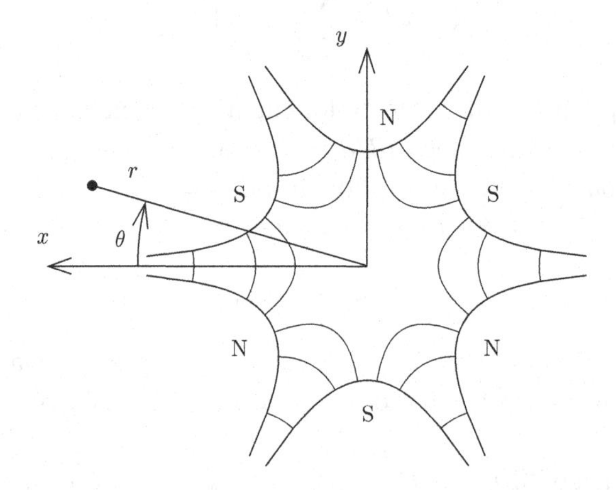

Figure 6.1 Polar co-ordinate system (r, θ) used for multipole expansion, superimposed on the field of an iron-dominated sextupole magnet. See also the right-handed co-ordinate system (x, y, s) introduced in Figure 2.2. The field lines are perpendicular to the contour of the pole when the relative permeability of the iron μ_r is very much greater than one.

and the strength of the magnetic field – the length of the field vector $\vec{B}$ – depends only on the radius

$$|\vec{B}| = b_n r^n \tag{6.8}$$

There are $2(n+1)$ north and south poles in an iron magnet of order n, although the unit vector $\widehat{B}$ describing the direction of the field

$$\widehat{B} = \sin(n\theta)\,\widehat{x} + \cos(n\theta)\,\widehat{y} \tag{6.9}$$

rotates only n times as its location rotates once around the origin. For example, $n = 2$ and

$$\begin{aligned} B_y &= b_2\, r^2 \cos(2\theta) = b_2\,(x^2 - y^2) \\ B_x &= b_2\, r^2 \sin(2\theta) \;= b_2\; 2xy \end{aligned} \tag{6.10}$$

in the case of the iron-dominated sextupole shown in Figure 6.1. This rotational symmetry explains why b_n and a_n are often called harmonic coefficients.

6.2 Iron-Dominated Magnets

How is a magnet designed to achieve a goal c_n?

The poles of an *iron-dominated* magnet are shaped to be perpendicular to the desired field shape, as illustrated in Figure 6.1 for a sextupole. This exploits the iron–vacuum boundary condition that applies when the relative permeability μ_r of the iron is much greater than the value of $\mu_r = 1$ in air or vacuum. The magnitude of c_n depends on how much flux enters (or leaves) each iron pole: on how many ampere-turns of current are carried by the conductors that are buried in the crevices between the poles.

This perpendicular boundary condition breaks down when the current is increased to the point where the iron starts to saturate, typically at values of 1 T or somewhat more, and μ_r is no longer very large in the iron. Saturated iron causes additional harmonics to become non-zero, but only some harmonics are allowed, depending on the geometric symmetries that are preserved even in saturation.

Up-down symmetry in a pure normal multipole guarantees that no a_n values are allowed, no matter how high the field. Similarly, m-fold rotational symmetry guarantees that b_n is only allowed if it drives Dm-fold distortions, where D is an odd number. For example, the allowed harmonics generated by a rotationally symmetric saturated sextupole are 18-pole and 30-pole, etc. Similarly, allowed harmonics for dipoles are 6-pole, 10-pole and 14-pole (et cetera), and for quadrupoles are 12-pole and 20-pole (et cetera).

6.3 Conductor-Dominated Magnets

Fields as high as 10 T or more are made possible by moving the iron as far away as possible, and by moving the conductor as close as possible, to the beam. Moving the iron out ensures that variable saturation (during an acceleration ramp) does not lead to unacceptable field quality at the beam. Iron is still important in providing a flux return-path that magnetically shields the outer environment, and in providing structural strength, for example to constrain conductor movement despite the very strong Lorentz forces. Very high current densities are required for an efficient magnet with a manageably small conductor area, so *conductor-dominated* is practically synonymous with *superconducting*.

Conductor shaped as a thin circular ring of radius a_0, with the ideal $\cos(n\theta)$ current distribution

$$J(r,\theta) = j_0\,\delta(r - a_0)\,\cos(n\theta) \tag{6.11}$$

generates a pure normal $2(n+1)$-pole field, with only b_n non-zero. A thick circular ring $\cos(\theta)$ distribution can be constructed, in the abstract, by two horizontally displaced cylinders of constant current density $|J_0|$, with opposite polarities, as shown in Figure 6.2. These current cylinders generate a perfect dipole field in the region where they overlap, even when the displacement Δ is large [34]. A practical realisation of a $\cos(\theta)$ magnet, the RHIC main arc dipole, is shown in cross-section in Figure 6.3.

A superconductor undergoes a phase transition and becomes normal conducting when any piece of it crosses a boundary surface in (B, J, T) space, where B is field, J is current density, and T is temperature. Hence it is the pole-tip field that best characterises the maximum performance of a multipole magnet, whether the real pole is nearing saturation in an iron-dominated magnet, or the euphemistic 'pole' at

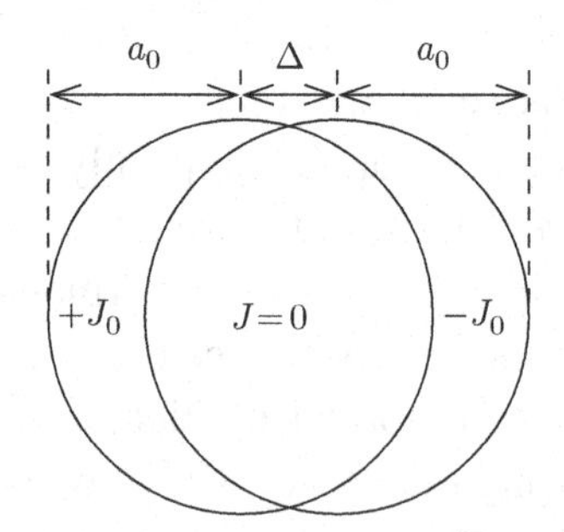

Figure 6.2 The net current distribution delivered by two overlapping current cylinders is close to the ideal $\cos(\theta)$ distribution described in Equation 6.11 (with $n = 1$) when the displacement Δ is small compared to the radius a_0. Even when Δ is large, this current distribution delivers a perfect dipole field in the overlap region.

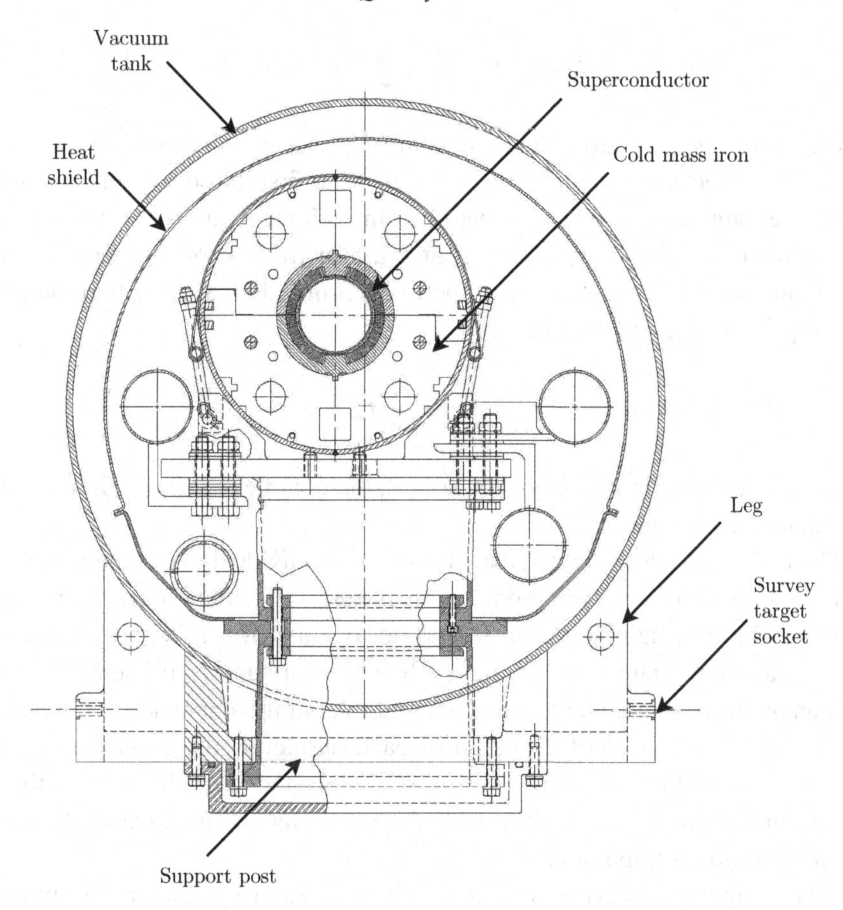

Figure 6.3 Cross-section of a RHIC main arc dipole in its cryostat. The superconductor carries a total of 5 kA in a series of blocks arranged with a cosine-like distribution just outside the beampipe aperture. (Courtesy of P. Wanderer.)

the conductor radius is quenching in a superconducting magnet. For example, if the superconductor in a quadrupole can support a maximum field of 8 T, the maximum field gradient

$$G_{max} = \frac{B_{max}}{a_0} \tag{6.12}$$

is about 200 T/m when the radial aperture is 0.04 m. Smaller apertures enable stronger magnets, except for dipoles.

6.4 Field Quality and Errors

The magnetic field quality of a set of magnets, all intended to be identical, is usefully discussed as an error relative to the goal field profile. For example, the actual field of a particular dipole is written as

$$B_y + iB_x = B_0(I)\left(1 + 10^{-4}\sum_{n=1}^{\infty}(b_n + ia_n)\left(\frac{x+iy}{R_0}\right)^n\right) \tag{6.13}$$

where there is potential for notational confusion, because the harmonic coefficients (b_n, a_n) introduced here are not the same as those first introduced in Equation 6.1. Here, they are conventionally quoted in dimensionless 'units' of 10^{-4}, at a reference radius R_0 that is sensibly set at a useful fraction of the magnet aperture that is intended to be a good-field aperture. Similarly, the error harmonics for a quadrupole are defined through

$$B_y + iB_x = G_0(I)\,x\left(1 + 10^{-4}\sum_{n=2}^{\infty}(b_n + ia_n)\left(\frac{x+iy}{R_0}\right)^{n-1}\right) \tag{6.14}$$

where the summation now begins at $n = 2$, the goal field is $B_y = G_0(I)x$, and I is the excitation current.

The harmonics (b_n, a_n) that characterise each individual magnet are often themselves also functions of excitation, I. This is not so important for accelerators (like most synchrotron light sources) that do not operate over a large dynamic range, but is vitally important for accelerators (like circular hadron colliders) that operate over an order-of-magnitude range of energies. A family of magnets is characterised by the mean and standard deviation of each harmonic, represented by $\langle b_n \rangle$ and σ_{bn} for many values of n. In practice, it is only the mean harmonics that vary significantly with excitation, often with allowed harmonics dominating the statistics when saturation is important.

Other important sources of error include random conductor misalignments, which drive random harmonics in conductor-dominated magnets. Systematic sources of construction or mechanical error can break the design symmetries, enabling non-allowed harmonics to acquire significant mean values. For example, perhaps the magnet is split in the mid-plane, or the distorting forces are asymmetric. Systematic harmonics often dominate random field errors in realistic magnet families [41].

Magnet field quality measurements are usually performed soon after each magnet comes off the production line, not long before it is installed in the tunnel, or is assembled into a composite magnet. Trends in mean harmonics may be noticed, needing correction in the production line. Often the availability of storage space limits the size of the pool of uninstalled (but available) magnets, and logistics limits the number of locations available for installation. Both of these limits hamper the ability to sort magnets by placing them at specific desirable locations, depending on their measured characteristics.

Installation misalignments also contribute to magnetic field errors. For example, a transversely displaced quadrupole leads to a dipole error, and an azimuthally

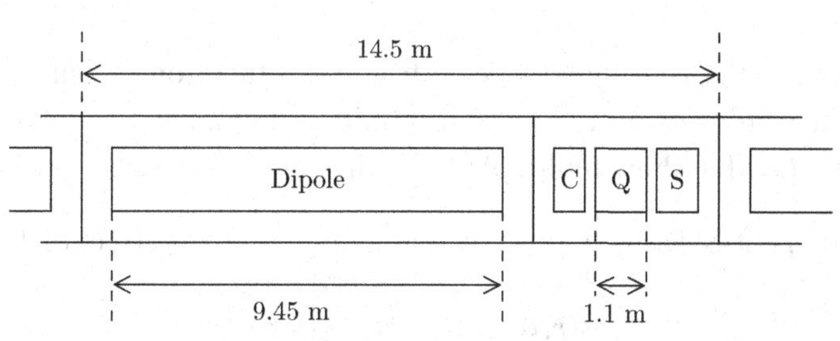

Figure 6.4 Superconducting magnets in two cryomodules forming a FODO half-cell in a RHIC main arc. One of the cryomodules contains a single dipole that is 9.45 m long. The other cryomodule contains a composite of three magnets: C is a corrector, Q is a quadrupole, and S is a sextupole. Some C correctors (such as those near the interaction point) have four concentric layers of independent correctors, while most are single-layer.

rotated normal quadrupole acquires a skew quadrupole component. While individual magnet misalignments can often be corrected in the tunnel, composite magnets can acquire irrevocable relative misalignments during construction and assembly. For example, Figure 6.4 illustrates how Corrector–Quadrupole–Sextupole composites are welded together from component magnets and then cryostatted, before installation in RHIC main arc FODO half-cells.

Exercises

6.1 Show that the expansion in Equation 6.1 solves Maxwell's equations.

6.2 Use conformal mapping to show that the contour of the poles of a quadrupole are given by

$$xy = \text{constant} \tag{6.15}$$

and that the poles of the sextupole shown in Figure 6.1 are given by

$$3x^2y - y^3 = \text{constant} \tag{6.16}$$

6.3 A 2-D magnetic field line follows a trajectory $(r(s), \theta(s))$, where s is the path length along the trajectory.

a) Given that

$$\frac{dr}{d\theta} = \frac{dr}{ds}\frac{ds}{d\theta} \tag{6.17}$$

show that

$$\frac{dr}{d\theta} = \frac{\widehat{B} \cdot \widehat{r}}{|\widehat{B} \times \widehat{r}|}\, r \tag{6.18}$$

b) Expand the right hand side to become a function of only b_n, r, θ and n, the order of a pure normal multipole field. This expression can be integrated to draw multipole field lines in (r, θ) space.

6.4 Use the Biot-Savart law to show that a thin circular current distribution

$$J(r, \theta) = j_0\, \delta(r - a_0)\, \cos(n\theta) \tag{6.19}$$

leads to a pure multipole inside a radius of a_0.

6.5 Show that the overlapping current cylinders of Figure 6.2 generate a perfect dipole field in the overlap region, for any value of displacement Δ that is less that $2a_0$, the diameter of each cylinder. Hint: use linear superposition.

6.6 Consider the window-frame dipole magnet shown in the figure, with gap width $w = 12$ cm, gap height $g = 2.5$ cm and length $l = 1$ m.

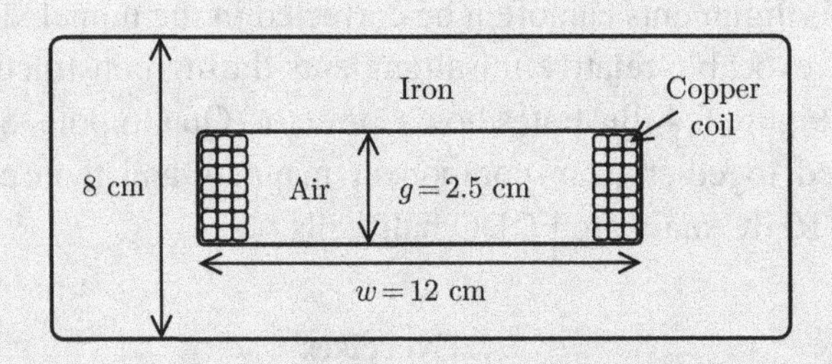

a) Show that the magnetic field in the gap is

$$B = \mu_0 \frac{NI}{g} \tag{6.20}$$

where N is the number of turns and I is the current in each turn, if the iron has infinite relative permeability $\mu_r \equiv (\mu/\mu_0) = \infty$.

b) Show that the inductance of the magnet is

$$L = \mu_0 N^2 \frac{lw}{g} \tag{6.21}$$

c) If the resistivity of copper is ρ, and the cross sectional area of the copper in each of the two coils is A, what is the total power dissipation in the magnet?

d) How many ampere-turns (NI) are necessary to achieve a field of 0.6 T in the gap, if the relative permeability of the iron is $\mu_r = 5000$. What, approximately, is the maximum field in the iron? Where?

e) Air-cooled copper coils carry a maximum current density of about 1.5 A/mm^2, while water-cooled copper coils can carry almost 10 times as much. However, water-cooling adds the potential for water leaks and is more expensive, so it is avoided if possible. Would you recommend water-cooled or air-cooled coils for this magnet? How much horizontal space is available between the coils for a beampipe?

f) If the magnet power supply has a maximum current of 1000 A, how many turns should be used in the coil?

g) Calculate the inductance of the magnet. How much energy is stored in the gap? How much power is dissipated in the coils?

h) Assuming constant field in the iron, estimate the additional energy stored in the iron yoke.

6.7 One of Maxwell's laws states that $\nabla \times \vec{B} = 0$ in a current-free region. In this case the field is described by scalar potential Φ where $\vec{B} = \nabla\Phi$.

a) If the horizontal and vertical fields in a quadrupole are $B_x = B'\,y$ and $B_y = B'\,x$, where B' is the constant field gradient, what is the shape of an equipotential surface?

b) Why is the pole-tip of an ideal quadrupole ($\mu_r = \infty$) an equipotential surface?

c) The radius of the inscribed circle touching the pole tips of the quadrupole is a. Show that the field gradient is

$$B' = \mu_0 \frac{NI}{a^2} \tag{6.22}$$

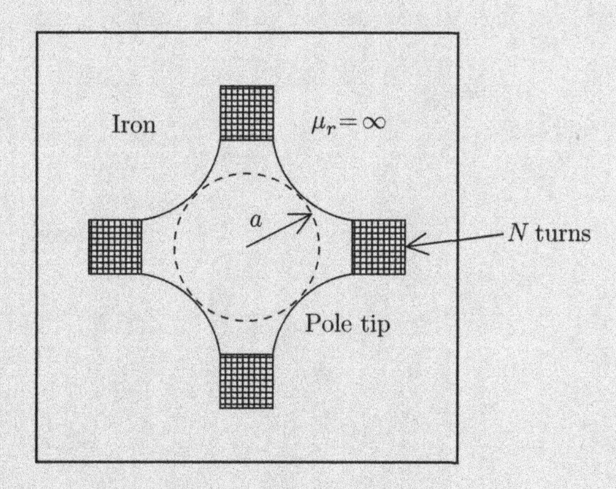

6.8 An antiproton beam passes from left to right through the lithium lens of length l and radius a that is shown in the figure. Interactions with lithium nuclei and atomic electrons are not significant. A total current I flows parallel (or antiparallel) to the beam, with a uniform distribution.

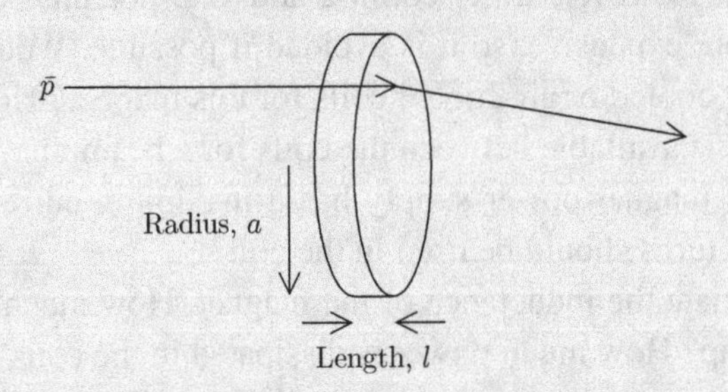

a) What is the focal length of the lens?
b) For focusing, must the current flow from left to right, or right to left?

6.9 Consider the general form for the scalar magnetic potential

$$\Phi = C r^n \sin(n\phi) \tag{6.23}$$

a) Show that Laplace's equation $\nabla^2\Phi = 0$ is satisfied.
b) Show that in cylindrical co-ordinates the magnetic field is

$$\begin{aligned} B_r &= C n r^{n-1} \sin(n\phi) \\ B_\phi &= C n r^{n-1} \cos(n\phi) \end{aligned} \tag{6.24}$$

c) If such an iron magnet is built, how many poles does it have?

7
RF Cavities

7.1 Waveguides

Consider an infinitely long open-ended waveguide with an unchanging cross-section, as sketched in Figure 7.1. If the walls are perfectly conducting, then the boundary conditions for oscillating fields are

$$\begin{aligned} \hat{n} \times \vec{E} &= 0 \\ \hat{n} \cdot \vec{B} &= 0 \end{aligned} \tag{7.1}$$

where $\hat{n}$ is a unit vector normal to the wall: the electric field is always perpendicular to the wall, and the magnetic field is always parallel to it. The fields within a waveguide driven by an RF power source at an angular frequency $\omega = 2\pi f$ oscillate, spatially and temporally, according to

$$\begin{aligned} \vec{E} &= \vec{E}(r,\theta)\, e^{ikz-\omega t} \\ \vec{B} &= \vec{B}(r,\theta)\, e^{ikz-\omega t} \end{aligned} \tag{7.2}$$

where r and θ are polar co-ordinates, z is the co-ordinate along the waveguide axis, and the real parts of the complex 3-D vectors $\vec{E}$ and $\vec{B}$ correspond to the physical fields. The wave number k labels a wave that propagates in the z-direction at the phase velocity

$$v_P = \frac{\omega}{k} \tag{7.3}$$

The effects of the boundary conditions at the walls are not trivially obvious. For example, k is imaginary under some conditions, in which case the wave is damped and fails to propagate long distances.

How does k vary as a function of ω? What happens when perfectly conducting walls are placed at both ends of a finite-length waveguide, turning it into a resonant cavity?

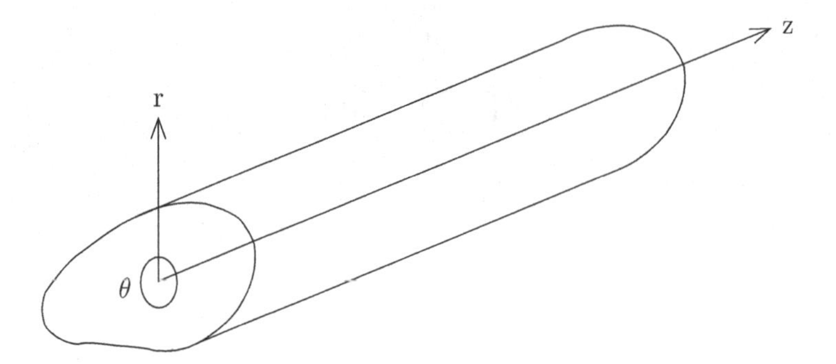

Figure 7.1 Polar co-ordinates in an infinitely long open-ended waveguide with an unchanging arbitrary cross-section.

7.2 Transverse Modes

It can be shown that there are three categories of solutions like Equation 7.2 that satisfy the boundary conditions of Equation 7.1 [21, 36, 57]:

1. TRANSVERSE MAGNETIC (TM) modes have
   $B_z = 0$ everywhere, with the boundary condition
   $E_z = 0$ at the walls.
2. TRANSVERSE ELECTRIC (TE) modes have
   $E_z = 0$ everywhere, with the boundary condition
   $\partial B_z/\partial n = 0$ at the walls.

TM modes are of most interest in accelerator design, since TE modes are incapable of accelerating or confining a beam. Nonetheless, TE modes are also important, for example in considering the higher-order modes that an RF power source or the beam itself can excite in a cavity, in addition to the fundamental TM mode. TE modes are occasionally useful when a transverse beam deflection that varies rapidly with time is required, for example in a crab cavity delivering a kick that varies with longitudinal displacement along a bunch.

In addition to the many TE and TM modes there is also a single degenerate Transverse ElectroMagnetic (TEM) mode, which has no longitudinal fields at all. This is familiar as the transverse wave that propagates in a medium with a wave number

$$k = \sqrt{\mu_R \epsilon_R}\, \frac{\omega}{c} \tag{7.4}$$

where μ_R and ϵ_R are the relative permeability and permittivity of the medium. Assuming from here on that $\mu_R = \epsilon_R = 1$ (in vacuum) then TEM-waves propagate at the speed of light, and are 'unaware' of the boundary walls because their wavelengths

$$\lambda = \frac{2\pi}{k} \tag{7.5}$$

are much smaller than the transverse dimensions of the waveguide.

Solving for TE or TM modes in a waveguide amounts to identifying a family of solutions that is labelled by a set of cut-off frequencies ω_n, each with a corresponding wave number

$$k_n = \frac{1}{c}\sqrt{\omega^2 - \omega_n^2} \tag{7.6}$$

Clearly k_n is imaginary and mode n is damped when the excitation frequency ω is less than the cut-off frequency for that mode. A stable mode with a real wave number propagates either forwards or backwards, depending on whether k is positive or negative.

At higher frequencies (and shorter wavelengths) each mode comes closer to behaving like a free-space TEM mode. Many propagating modes can co-exist, as illustrated in Figure 7.2, but there is a range of excitation frequencies $\omega_0 < \omega < \omega_1$

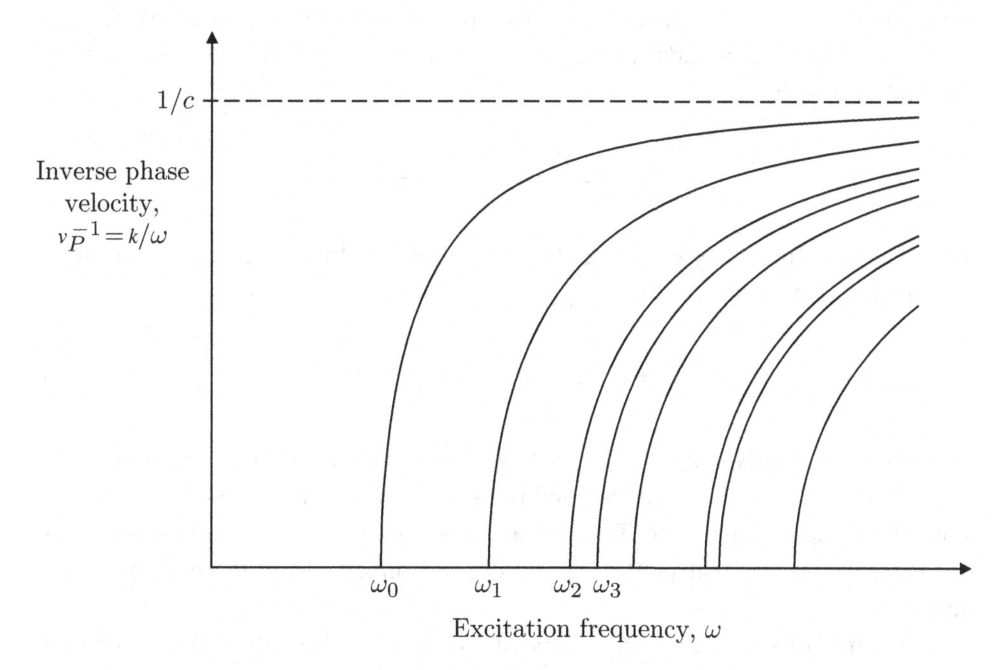

Figure 7.2 The dependence of wave number k on excitation frequency ω for a family of TE or TM waveguide modes with cut-off frequencies ω_n.

in which only the lowest frequency fundamental mode is stable. It is desirable for $\omega_1 - \omega_0$, the difference between the fundamental and the first higher-order mode cut-off frequencies, to be large.

The beam itself excites modes as it passes through a waveguide or cavity, even if the drive frequency is carefully separated from higher-order modes. A train of bunches each with an RMS length σ_z can couple (strongly or weakly) to modes with frequencies as high as about

$$\omega_{MAX} \approx 2\pi \frac{c}{\sigma_z} \tag{7.7}$$

Bunches can excite many higher-order modes if they are short compared to the transverse dimensions of a waveguide, or short compared to the longitudinal features of a cavity!

7.3 Cylindrical Resonant Cavities – Pill-Boxes

Cavities are more interesting than waveguides, so consider now a waveguide of arbitrary cross-section that is closed with flat ends at $z = 0$ and $z = L$. Cavity mode solutions that respect the boundary conditions of Equation 7.1 on the end walls are readily constructed by combining pairs of forward and backward propagating modes with equal amplitudes and appropriate phases. A pair of waveguide modes labelled by p with

$$k = \pm k_p = \pm\frac{p\pi}{L}, \qquad p = 0, 1, 2, ... \tag{7.8}$$

have the same frequency ω, and add (for example) to form a resonant TM mode with a natural resonant frequency ω_{RES}

$$E_z = \psi(r,\theta) \cos\left(p\pi \frac{z}{L}\right) e^{-i\omega_{RES} t} \tag{7.9}$$

This mode explicitly respects the end wall boundary conditions, while $\psi(r,\theta)$ respects the side wall boundary conditions for a waveguide. A similar expression (in B_z) applies for each TE resonant mode p. Transverse field components for TM and TE resonant modes can be derived directly from these longitudinal fields [21].

The family of solutions $\psi(r,\theta)$ depends on the cross-section of the structure – square, rectangular, circular or complicated and realistic. Pill-box geometry is instructive: a cylindrical cavity of radius R and length L has a complete set of TM modes that is labelled by three indices mnp, through

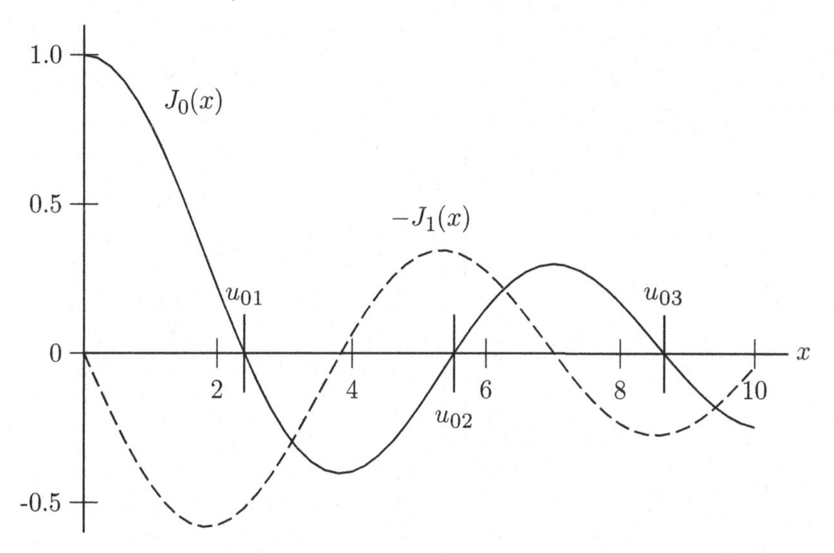

Figure 7.3 The first-order Bessel function $J_0(x)$ and its slope $J_0'(x) = -J_1(x)$. The first three roots are at $(u_{01}, u_{02}, u_{03}) = (2.405, 5.520, 8.654)$.

$$
\begin{aligned}
E_z &= E_0 \cos\left(p\pi\frac{z}{L}\right) \cdot J_m\left(u_{mn}\frac{r}{R}\right) \cdot \cos(m\theta) \cdot e^{-i\omega_{mnp}t} \\
E_r &= -E_0 \frac{p\pi R}{u_{mn}L} \sin\left(p\pi\frac{z}{L}\right) \cdot J_m'\left(u_{mn}\frac{r}{R}\right) \cdot \cos(m\theta) \cdot e^{-i\omega_{mnp}t} \\
E_\theta &= E_0 \frac{mp\pi R}{u_{mn}^2 L} \sin\left(p\pi\frac{z}{L}\right) \cdot \frac{R}{r} J_m\left(u_{mn}\frac{r}{R}\right) \cdot \sin(m\theta) \cdot e^{-i\omega_{mnp}t} \\
B_z &= 0 \\
B_r &= B_0 \frac{m\omega_{mnp}R}{u_{mn}^2 c} \cos\left(p\pi\frac{z}{L}\right) \cdot \frac{R}{r} J_m\left(u_{mn}\frac{r}{R}\right) \cdot \sin(m\theta) \cdot e^{-i\omega_{mnp}t} \\
B_\theta &= B_0 \frac{\omega_{mnp}R}{u_{mn}c} \cos\left(p\pi\frac{z}{L}\right) \cdot J_m'\left(u_{mn}\frac{r}{R}\right) \cdot \cos(m\theta) \cdot e^{-i\omega_{mnp}t}
\end{aligned}
\tag{7.10}
$$

where J_m is the mth Bessel function of the first kind, J_m' is the differential of J_m and u_{mn} is the nth root of the mth Bessel function. For example, Bessel function J_0, its slope and roots are shown in Figure 7.3. The natural resonant frequency of TM mode mnp is

$$
\omega_{mnp} = c\sqrt{\left(\frac{u_{mn}}{R}\right)^2 + \left(\frac{p\pi}{L}\right)^2}
\tag{7.11}
$$

showing that accelerating modes with $p = 0$ have the lowest frequencies.

Simplifying further, the accelerating modes with no azimuthal structure $m = p = 0$ have only two non-zero field components

$$
\begin{aligned}
E_z &= E_0 \qquad J_0\left(u_{0n}\frac{r}{R}\right) e^{-i\omega_{0n0}t} \\
B_\theta &= -B_0 \frac{\omega_{0n0}R}{u_{0n}c} J_1\left(u_{0n}\frac{r}{R}\right) e^{-i\omega_{0n0}t}
\end{aligned}
\tag{7.12}
$$

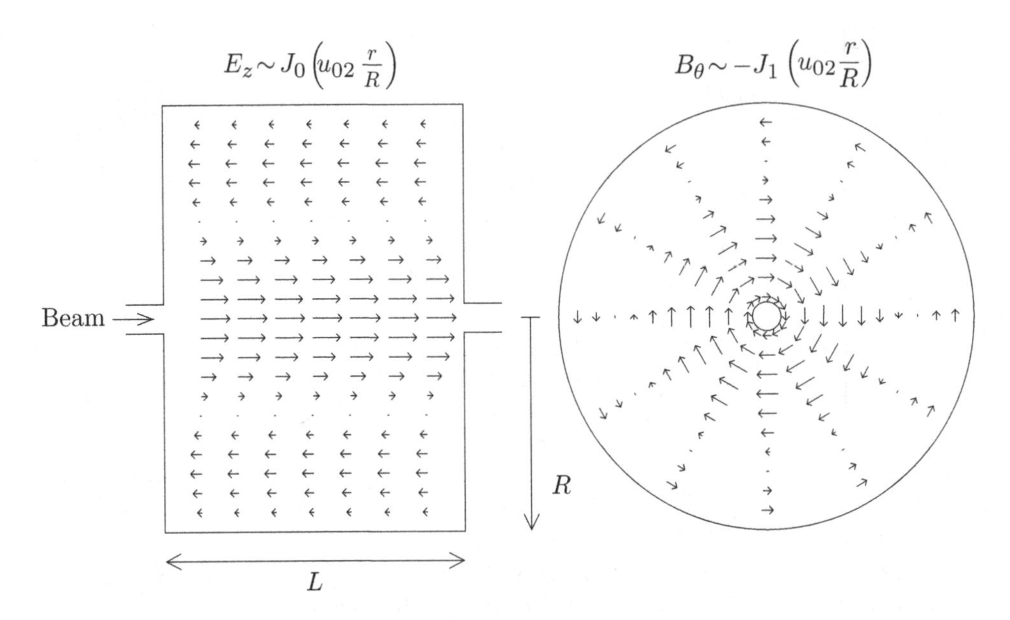

Figure 7.4 Longitudinal and azimuthal fields for the transverse magnetic mode TM_{020} in a pill-box cavity.

where the identity

$$J_0'(x) = -J_1(x) \tag{7.13}$$

is used. Field patterns for the TM_{020} mode are shown in Figure 7.4, illustrating how $n = 2$ counts the number of zero-crossings that E_z encounters between the axis and the radial wall. Similarly, $m = 0$ counts the number of periods in the azimuthal structure, while $p = 0$ counts the number of zero-crossings between the two end walls. Note that the magnetic field on the wall $B_\theta(R)$ is non-zero, since

$$J_1(u_{0n}) \neq 0 \tag{7.14}$$

A cavity is not much use unless beam can travel through it, and so beampipe holes are cut on the axis at both ends. Fortunately the holes do not perturb small-n modes very much, if their radius is much smaller than R. This can be said a different way: a mode is not much perturbed if its resonant frequency

$$\omega_{0n0} = \frac{u_{0n}c}{R} \geq 2.405\,\frac{c}{R} \tag{7.15}$$

is much smaller than the cut-off frequency of the beampipe, so fields leak only a little way into the beampipe. The fundamental accelerating mode leaks far enough to allow an input power coupler to be placed there. If necessary, higher-order mode couplers may also be placed in the beampipe just outside the cavity, extracting energy from modes that can cause beam instabilities. See Section 13.3 for more discussion of the beampipe interface.

7.4 Cavity Performance Limits

The accelerating field changes with time as a particle goes from end to end in a single cavity. What is the maximum energy eV_A that it can gain during passage? What is the optimum length of the cavity?

A particle with speed βc that passes the middle of a cavity at time $t = 0$ acquires an accelerating voltage

$$V_A = \int_{-L/2}^{L/2} E_z \, dz = \beta c \, E_0 \int_{-L/(2\beta c)}^{L/(2\beta c)} e^{i\omega t} \, dt \tag{7.16}$$

from a TM_{0n0} mode, so that

$$V_A = E_0 L \cdot T_1 \tag{7.17}$$

The single-cell transit-time factor T_1 depends on the cavity length L through

$$T_1(L) = \frac{\sin(\omega L/2\beta c)}{\omega L/2\beta c} \tag{7.18}$$

and so V_A has a maximum when $T_1 = 2/\pi$ with an optimum length

$$L_{opt} = \pi \frac{\beta c}{\omega} \tag{7.19}$$

For example, a pill-box operating in the TM_{010} mode has an optimum length

$$L_{opt} = \frac{\pi}{2.405} \beta R \tag{7.20}$$

that depends on the speed of the particle: pill-boxes become unacceptably inefficient for low-energy non-relativistic hadrons with $\beta \ll 1$.

Cavity geometries far from pill-box are necessary for non-relativistic particles, typically when $\beta \leq 0.5$. The cavity zoo includes relatively exotic creatures like spoke, split-ring, split-loop resonators and inter-digital structures, in addition to relatively mundane quarter-wave and half-wave resonators. Cavities designed to operate at relativistic velocities morph more gently away from the single-cell pill-box geometry, for example into the single-cell and multi-cell elliptical cavities illustrated in Figure 7.5. The language of TM_{mnp} modes survives the morph, with the indices still counting zero-crossings and azimuthal periods, because the topology and the rotational symmetry are still fundamentally the same.

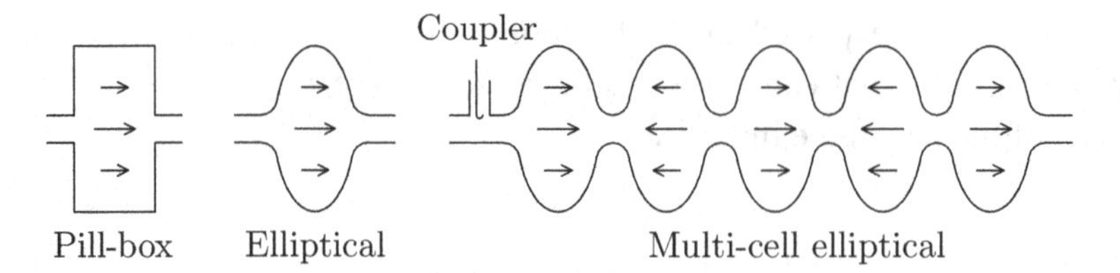

Figure 7.5 Single-cell and multi-cell RF cavity topologies. The modes in a single-cell elliptical cavity are very similar to those in a pill-box. The fundamental accelerating TM mode in a multi-cell cavity includes a π phase advance from cell to cell.

Multi-cell cavities overcome the maximum-length limit imposed by the single-cell transit-time factor, invoking a π phase advance between cells to ensure that a particle receives (near) optimum acceleration throughout the cavity. The geometry and alignment of a multi-cell cavity must be sufficiently accurate to ensure that the field profile of the fundamental mode is reasonably flat from cell to cell, and to suppress or move beam-driven higher-order modes. See Section 13.2 for more discussion of multi-cell cavities.

Field strengths in room-temperature copper cavities are limited by the maximum electric field E_k that is possible at the walls, before surface breakdown occurs. The empirical Kilpatrick criterion

$$f = 1.64\,E_k^2 \exp\left(\frac{-8.5}{E_k}\right) \tag{7.21}$$

(where f is in MHz, and E_k is in MV/m) shows that higher frequencies enable higher wall fields, with values in the range from 11 MV/m to 47 MV/m for typical frequencies in the range from 100 MHz to 3 GHz. Modern vacuum and surface preparation practices make this estimate rather conservative – the maximum estimated field can be increased by bravery factors of as much as 2, with some confidence.

Superconducting operation is possible for most cavity geometries. Superconducting cavities have the advantage of providing higher performance (with accelerating gradients sometimes much greater than 10 MV/m) at lower operating cost (almost eliminating heat dissipation in resistive cavity walls). Very high fields and large Q-values of order 10^{10} together introduce a set of issues that limit superconducting cavity performance. For example, small mechanical vibrations (microphonics) and Lorentz force detuning can distort the cavity shape as a function of time, detuning the cavity by moving the resonant frequency by a distance comparable to the small value ω/Q, the characteristic bandwidth of the system. Field emission and thermal breakdown can cause the cavity to quench and become normal-conducting.

Despite these challenges, superconducting RF technology is at a frontier that is steadily advancing, sometimes quickly. Fortunately this book does not attempt to describe the current frontier in detail, and so should not need frequent revision.

Exercises

7.1 What is the ratio of the complex numbers B_0 and E_0 in Equation 7.10? Hint: see Equation 13.20.

7.2 Why is the TM_{010} mode usually preferred?

7.3 Shunt impedance is a figure of merit that evaluates the square of the voltage gain of a particle passing through a cavity, per unit of power loss to the surfaces of the cavity. One definition (of many) for a pill-box cavity of length L and radius R gives a shunt impedance

$$R_s = \frac{Z_0^2}{\pi\rho}\frac{L}{R}\frac{T_1^2}{(1+(R/L))J_1^2(2.4050)} \tag{7.22}$$

where T_1 is the transit-time factor and ρ is the surface resistivity.

a) What value of R/L maximizes the shunt impedance R_s?
b) If instead the shunt impedance per unit length $r_s = R_s/L$ is to be maximized, what is the optimum value of R/L?

7.4 The Kilpatrick criterion

$$f = 1.64\, E_k^2 \exp(-8.5/E_k). \tag{7.23}$$

is an empirical equation from the 1950s that predicts the relation between frequency f (in MHz) and electrical field E_k (in MV/m) on a room-temperature copper surface at the limit of electrical breakdown. Higher frequencies support higher gradients. Contemporary vacuum systems allow the Kilpatrick limit E_k to be exceeded by bravery factors as large as 2.

If the maximum surface field on the walls of a single-cell pill-box cavity is $1.8E_k$, then how many cavities are required to accelerate beam at 5 MeV per turn when the frequency is 200 MHz, 400 MHz and 800 MHz?

7.5 The figure shows a betatron, an early type of accelerator. Very low-energy electrons are injected into the periphery between two cylindrical magnet poles that drive a vertical magnetic field B which increases uniformly from a non-zero value, constraining the electrons to move (approximately) in a circle. A circular electric field $E \sim dB/dt$ accelerates the electrons. Most electrons deviate from the ideal orbit, depending on initial conditions.

a) What are perfect initial injection conditions?
b) What is the equation of motion of a perfect non-relativistic electron?
c) How close to perfect must the initial conditions be, in order for the electron to accelerate successfully?

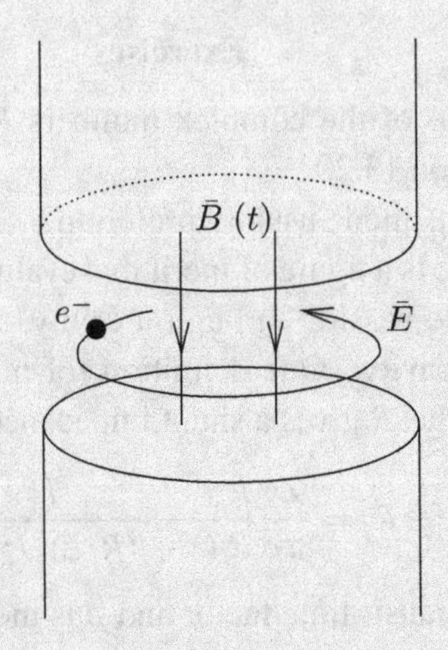

d) What (approximate) maximum energy can a betatron deliver? Why?

(This question can be answered analytically, in simulation, or both.)

8

Linear Errors and Their Correction

The linear errors discussed here – their sources, effects and correction – correspond to the first three correction magnets listed in Table 8.1. Later chapters discuss nonlinear correctors, like sextupoles and octupoles, that are necessary to control the accelerator dynamics, even in the absence of errors.

Linear errors and nonlinear fields (intentional or otherwise) lead to the need to avoid resonances: values of the horizontal and vertical tunes that come close to satisfying the relationship

$$p = q Q_x + r Q_y \tag{8.1}$$

where p, q and r are integers. Since the number line is everywhere dense in rational fractions, it is fortunate that only relatively small values of $|q|$ and $|r|$ are important. For example, the linear errors discussed here only drive resonances with $(q, r) = (1, -1), (1, 0), (0, 1), (2, 0)$ and $(0, 2)$. In contrast, p can be large, since Q_x and Q_y often have large integer components. Usually only the fractional parts of the tunes matter, since their adjustment alone is sufficient to avoid all important resonances.

8.1 Trajectory and Closed Orbit Errors

If a pure multipole magnet of strength c_n is accidentally offset transversely by Δx and Δy, then a linear expansion of the local magnetic field

$$\begin{aligned} B_y + iB_x &= c_n \left((x - \Delta x) + i(y - \Delta y)\right)^n \\ &= c_n \left[(x + iy)^n - n(x + iy)^{n-1}(\Delta x + i\Delta y) + \cdots\right] \end{aligned} \tag{8.2}$$

shows the phenomenon of multipole feed-down, through which the magnet acquires an error field of multipole order $n - 1$, with a first-order strength

$$c_{n-1} = -c_n \, n \, (\Delta x + i\Delta y) \tag{8.3}$$

Table 8.1 *Common correction magnets, and their applications. Dipoles, normal quadrupoles and skew quadrupoles correct linear errors that are inevitably present. Nonlinear magnets like sextupoles and octupoles are necessary even in a perfect accelerator, for example to set chromaticities and detuning coefficients to their desired values, or to drive slow extraction.*

Multipole	Magnet style	Correction applications
a_0, b_0	V & H dipoles	Closed orbits, injection
b_1	Normal quadrupole	Tunes, β-functions, transition
a_1	Skew quadrupole	Transverse decoupling
b_2	Sextupole	Chromaticities
b_3	Octupole	Tunes versus amplitude
$a_2, b_4, \ldots$	Skew octupole, decapole ...	Interaction region optics

that is linear in the misalignment. The higher-order feed-down to multipoles $n-2$, etc, can be ignored if Δx and Δy are small.

In particular, a thin quadrupole of integrated strength $q = 1/f$ that is offset by Δx delivers a total angular kick of

$$\Delta x' = -\frac{x}{f} + \frac{\Delta x}{f} \tag{8.4}$$

to a test particle with a betatron displacement x, including a constant dipole error kick of $\Delta x/f$ that is delivered on every passage. For example, a quadrupole with a focal length of $f = 10$ m that is misaligned by 1 mm delivers an angular kick of 100 μrad.

Transverse displacements of dipoles do not generate any angular kick errors. However, a dipole of horizontal bend angle θ that is accidentally rolled by a small angle α about the longitudinal axis delivers a vertical error kick of

$$\Delta y' = \alpha\theta \tag{8.5}$$

For example, a dipole with a 40 mrad bend angle that is accidentally rolled by 1 mrad delivers a vertical kick of 40 μrad.

Quadrupole offsets and dipole rotations are the major sources of dipole error kicks that distort trajectories and closed orbits. How big are these distortions, and how are they corrected?

Suppose that a particle with no initial betatron oscillation encounters a horizontal dipole error kick $\Delta x'$ at a location s_0. A free oscillation wave propagates downstream, with a horizontal displacement

$$\frac{x}{\sqrt{\beta}} = \Delta x' \sqrt{\beta_0}\, \sin(\psi(s) - \psi_0) \tag{8.6}$$

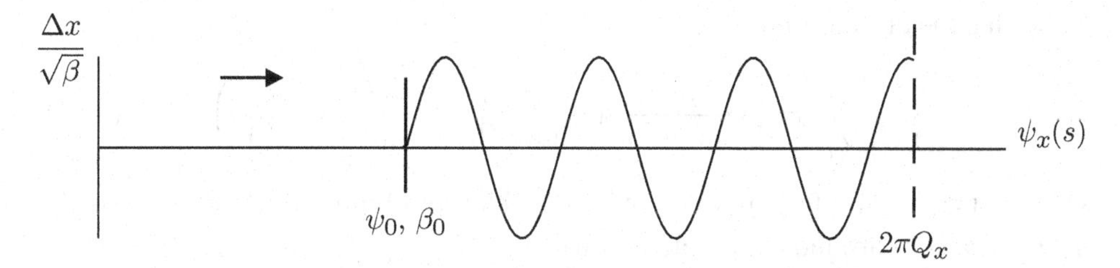

Figure 8.1 The free wave of the trajectory propagating downstream of an angular error kick at $\psi_x(s_0) = \psi_0$.

at a general location with $s > s_0$, according to the discussion in Section 3.3. For example, one 100 μrad kick launches a 5 mm oscillation in a FODO channel with typical β-values of 50 m. Not many such errors can be tolerated, without correction, before the beam encounters the beampipe.

A free wave does not reproduce itself when it returns after one complete turn, as illustrated in Figure 8.1. The closed orbit *does* repeat itself, by definition: periodic boundary conditions must be enforced. Thus, the closed orbit displacement at s_0 is found by solving the equation

$$\begin{pmatrix} x \\ x' \end{pmatrix}_{co} = M \begin{pmatrix} x \\ x' \end{pmatrix}_{co} + \begin{pmatrix} 0 \\ \Delta x' \end{pmatrix} \tag{8.7}$$

where M is the one-turn matrix, so that

$$\begin{pmatrix} x \\ x' \end{pmatrix}_{co} = (I - M)^{-1} \begin{pmatrix} 0 \\ \Delta x' \end{pmatrix} \tag{8.8}$$

where I is the identity matrix. Trouble is anticipated when $\det(I - M)$ is small!

Suppose for simplicity that the Twiss function $\alpha_0 = 0$, so that

$$I - M = \begin{pmatrix} 1 - C & -\beta S \\ S/\beta & 1 - C \end{pmatrix} \tag{8.9}$$

where

$$C = \cos(2\pi Q_x) \quad \text{and} \quad S = \sin(2\pi Q_x) \tag{8.10}$$

so that

$$(I - M)^{-1} = \frac{1}{2(1 - C)} \begin{pmatrix} 1 - C & \beta S \\ -S/\beta & 1 - C \end{pmatrix} \tag{8.11}$$

Using the trigonometrical identities

$$\begin{aligned} 1 - C &= 2\sin^2(\pi Q_x) \\ S &= 2\sin(\pi Q_x)\cos(\pi Q_x) \end{aligned} \tag{8.12}$$

shows that the closed orbit at s_0 is

$$\begin{pmatrix} x \\ x' \end{pmatrix}_{co} = \frac{\Delta x'}{2\sin(\pi Q_x)} \begin{pmatrix} \beta_0 \cos(\pi Q_x) \\ \sin(\pi Q_x) - \alpha_0 \cos(\pi Q_x) \end{pmatrix} \tag{8.13}$$

which in fact is valid for any value of α_0. The closed orbit wave caused by a single error propagates around the accelerator like

$$\frac{x_{co}}{\sqrt{\beta}} = \frac{\Delta x' \sqrt{\beta_0}}{2\sin(\pi Q_x)} \cos(|\psi - \psi_0| - \pi Q_x) \tag{8.14}$$

with a cusp at s_0 across which the closed orbit displacement remains constant, while the closed orbit angle increases by $\Delta x'$. This is illustrated in Figure 8.2. The closed orbit becomes extremely sensitive to dipole kick errors when the resonance denominator in Equation 8.14 is close to zero: that is, when

$$p \approx Q_x \tag{8.15}$$

an integer resonance with $(q, r) = (1, 0)$.

Closed orbit errors are corrected by dipole correctors, which are usually conveniently placed close to quadrupoles. Any three dipole correctors $i = 1, 2, 3$ perturb

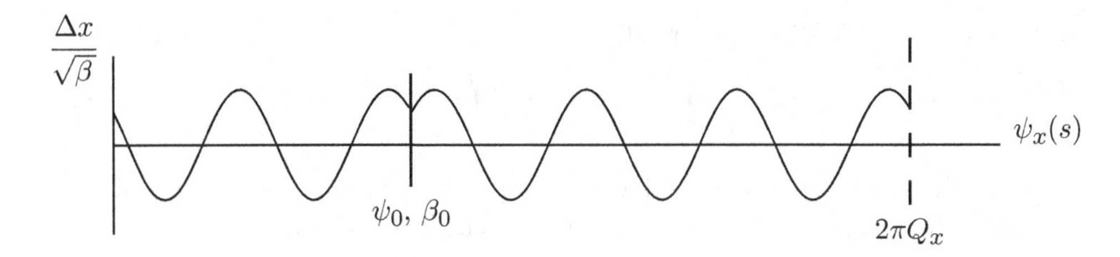

Figure 8.2 The closed orbit error wave due to an angular kick at ψ_0, after the application of periodic boundary conditions in an accelerator with tune $Q_x = 5.30$.

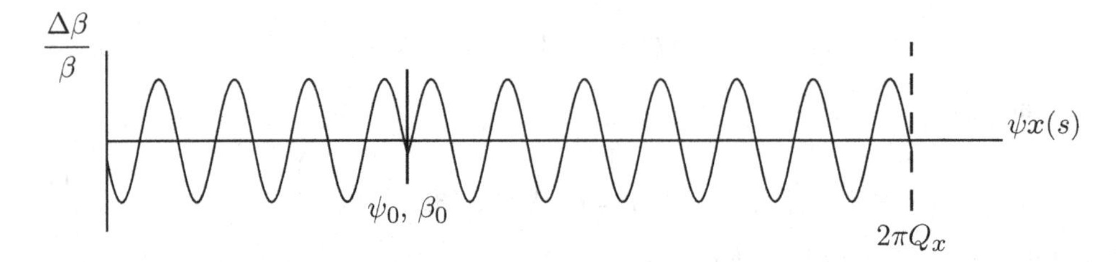

Figure 8.3 The beta wave due to a quadrupole perturbation error at ψ_0, propagating at twice the speed of trajectory and closed orbit error waves, in an accelerator with tune $Q_x = 5.30$.

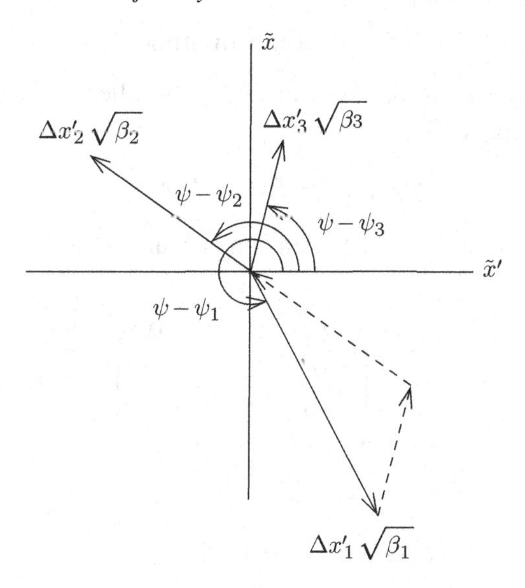

Figure 8.4 Three-bump vectors rotating in normalised phase space. The dashed lines illustrate how the vector sum of the three angular kicks add up to zero outside the three-bump.

the closed orbit only in the local region between the first and the third if the (free wave) oscillation downstream of all three sums to zero

$$\frac{x(\psi)}{\sqrt{\beta}} = \sum_{i=1}^{3} \Delta x'_i \sqrt{\beta_i} \sin(\psi - \psi_i) = 0 \tag{8.16}$$

It is easily shown that such three-bump localisation is guaranteed if

$$\frac{\Delta x'_i \sqrt{\beta_i}}{\sin(\psi_j - \psi_k)} = \kappa \tag{8.17}$$

where (i, j, k) is any cyclic combination of (1, 2, 3), and κ is the strength of the three-bump, as illustrated in Figure 8.4.

Closed orbit offsets are measured by beam position monitors that detect and interpret the electromagnetic signals generated by passing bunches. Each three-bump strength κ can be adjusted to minimise the sum of squares of the measured closed orbit displacements within the bump. This can be repeated in overlapping three-bumps – using dipole corrector triads (1,2,3), then (2,3,4), etc – to correct the closed orbit around the entire ring, or only in any desired segment. This overlapping three-bump method is only one of many algorithms that are routinely applied to closed orbit correction. In practice, repeated iterations are necessary, in calculation and in application, before closed orbits are optimised.

8.2 Linear Coupling

The matrix for a magnet that has been accidentally rolled clockwise by an angle α about the longitudinal axis is

$$M_{ROLLED} = R(-\alpha)\, M\, R(\alpha) \tag{8.18}$$

where M is the matrix of the unrolled magnet and R is the co-ordinate transformation

$$R = \begin{pmatrix} C & 0 & S & 0 \\ 0 & C & 0 & S \\ -S & 0 & C & 0 \\ 0 & -S & 0 & C \end{pmatrix} \tag{8.19}$$

where

$$C = \cos\alpha \quad \text{and} \quad S = \sin\alpha \tag{8.20}$$

The rolled thin quadrupole matrix is therefore

$$M_{RTQ}(\alpha) \approx M_{TQ} + \alpha \left(\begin{array}{cc|cc} 0 & 0 & 0 & 0 \\ 0 & 0 & -2q & 0 \\ \hline 0 & 0 & 0 & 0 \\ -2q & 0 & 0 & 0 \end{array}\right) \tag{8.21}$$

to first order in α, where M_{TQ} is the matrix for a thin normal quadrupole (see Equation 2.26). This matrix couples horizontal and vertical motion because it is not block diagonal, unlike the matrices (like M_{TQ}) that represent ideal dipoles and quadrupoles.

There is a strong analogy between coupled motion in an accelerator and the weakly coupled motion of the two pendula shown in Figure 8.5. If there is no coupling (when the spring constant $k = 0$) each pendulum has a unique characteristic frequency for small oscillations, scaling with its length according to

$$f \sim \frac{1}{\sqrt{L}} \tag{8.22}$$

When k is non-zero, the evolution of the two angles θ_1 and θ_2 depends on two eigenfrequencies, f_A and f_B, according to

$$\begin{aligned} \theta_1 &= a_{1A} \cos(f_A t + \phi_{1A}) + a_{1B} \cos(f_B t + \phi_{1B}) \\ \theta_2 &= a_{2A} \cos(f_A t + \phi_{2A}) + a_{2B} \cos(f_B t + \phi_{2B}) \end{aligned} \tag{8.23}$$

where the relative sizes of the amplitude coefficients (a_{1A}/a_{1B} and a_{2B}/a_{2A}) reflect the influence of the coupling.

The pendula eigenfrequencies change as the length L_1 is varied, moving the uncoupled frequency f_1 across the natural uncoupled value f_2, as shown in the plot on the right of Figure 8.5. When L_1 is big and f_1 is small, then $f_B \approx f_1$ and $f_A \approx f_2$.

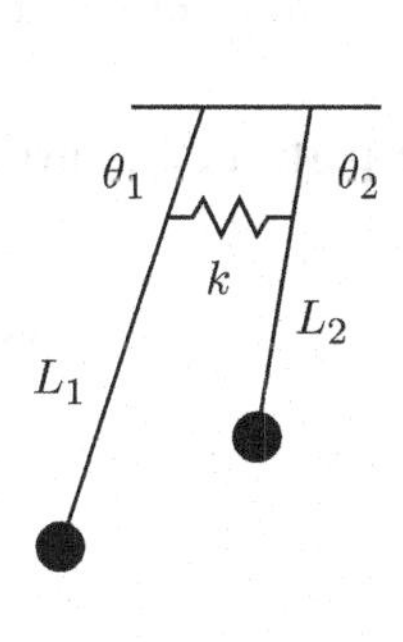

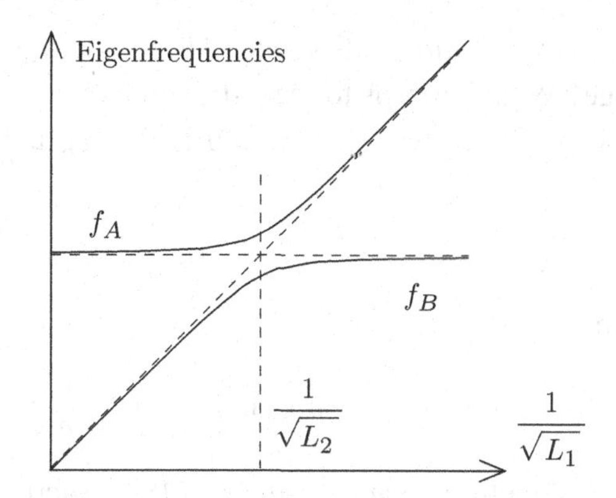

Figure 8.5 Two pendula, coupled by a weak spring with a restoring force of $k(\theta_1 - \theta_2)$. The two eigenfrequencies are close to the natural pendulum frequencies (which scale like $1/\sqrt{L}$) when the two lengths L_1 and L_2 are far apart. They interfere without crossing when $L_1 \approx L_2$.

Since the motion is close to uncoupled, then $a_{1A} \ll a_{1B}$ and $a_{2B} \ll a_{2A}$. In all cases $f_A > f_B$, so that at the other extreme, when L_1 is much shorter than L_2, then $f_B \approx f_2$ and $f_A \approx f_1$. In between, when $L_1 \approx L_2$, then the motion is strongly coupled, so that $a_{1A} \approx a_{1B}$ and $a_{2A} \approx a_{2B}$. How close L_1 and L_2 have to be for strong coupling depends on the strength of the spring constant k.

The natural (uncoupled) pendulum frequencies f_1 and f_2 are analogous to the design values of the accelerator tunes Q_x and Q_y. The spring constant k is analogous to ΔQ_{min}, the closest approach of the accelerator eigentunes, which measures the amount of coupling introduced by accidentally rolled quadrupoles and by any solenoidal fields. Horizontal and vertical motion is fully coupled and confused when the *design* fractional tunes are close enough to satisfy

$$|p + Q_x - Q_y| \lesssim \Delta Q_{min} \tag{8.24}$$

the coupling resonance condition with $(q, r) = (1, -1)$. Usually ΔQ_{min} is much less than one, and the integer p absorbs the integer difference of the design tunes. This reflects, once again, the nature of an accelerator as a difference system, in contrast to the differential nature of gravity pendula.

Skew quadrupole correctors reduce the closest approach of the fractional tunes to values as small as $\Delta Q_{min} \approx 0.001$, from uncorrected values that can be two orders of magnitude larger, depending on the situation [14, 39].

8.3 Tune Shifts and β-Waves

If a quadrupole at s_0 has a strength error

$$\Delta q = \Delta \left(\frac{1}{f}\right) \tag{8.25}$$

is the linear motion still stable? How much do the horizontal and vertical tunes change? What happens to the β-functions?

The perturbed tune $\tilde{Q}$ is found by taking the trace of the perturbed matrix $\tilde{M}$, since

$$\mathrm{Tr}(\tilde{M}) = 2\cos(2\pi\tilde{Q}) \tag{8.26}$$

where

$$\tilde{M} = M\begin{pmatrix} 1 & 0 \\ -\Delta q & 1 \end{pmatrix} \tag{8.27}$$

in a thin quadrupole approximation. It is exactly true that

$$\cos(2\pi\tilde{Q}) = \cos(2\pi Q) - \frac{\beta_0\Delta q}{2}\sin(2\pi Q) \tag{8.28}$$

where β_0 is the unperturbed β-function at the quadrupole error, in the plane of interest. For small perturbations the linearised tune shift

$$\Delta Q = \tilde{Q} - Q = \frac{\beta_0\,\Delta q}{4\pi} \tag{8.29}$$

is independent of Q. In the other plane Δq changes sign, so the horizontal and vertical tunes move in different directions. For example, a relative error of 1% in a single quadrupole of focal length 10 m at a location in a FODO channel with $(\beta_x, \beta_y) = (50, 10)$ m leads to tune shifts of approximately $(\Delta Q_x, \Delta Q_y) = (0.0040, -0.0008)$. Fortunately it is relatively easy to correct tune perturbations, no matter whether quadrupole family strength errors are systematic or random, by adjusting the strengths of two families of quadrupoles in the main arcs.

Quadrupole errors also perturb the β-functions downstream of an error in a transfer line (in a free oscillation), or all around a circular accelerator (with periodic boundary conditions). The Twiss functions just downstream of a thin quadrupole error, found by applying Equation 3.34, are exactly

$$\begin{pmatrix} \tilde{\beta} \\ \tilde{\alpha} \\ \tilde{\gamma} \end{pmatrix}_{0+\epsilon} = \begin{pmatrix} 1 & 0 & 0 \\ \Delta q & 1 & 0 \\ (\Delta q)^2 & 2\Delta q & 1 \end{pmatrix} \begin{pmatrix} \beta \\ \alpha \\ \gamma \end{pmatrix}_0 \tag{8.30}$$

showing that β is unchanged, but the slope of β is perturbed to become

$$\tilde{\beta}' = -2\tilde{\alpha} = \beta_0' - 2\beta_0\Delta q \tag{8.31}$$

This launches a β-wave that propagates in a free oscillation according to

$$\frac{\Delta\beta}{\beta} = -\Delta q\beta_0\,\sin(2(\psi - \psi_0)) \tag{8.32}$$

advancing at *twice* the rate of the betatron phase advance. For example, a relative error of 1% in a single quadrupole with a focal length of 10 m at a location with $\beta_0 = 50$ m launches a (free) β-wave with a relative amplitude of 5%. Errors of this size can easily be fixed if they are due to systematic errors in a family of quadrupoles, but if they are random (and unknown *a priori*) they are very dangerous. As a rule of thumb, quadrupole strengths must be known and controlled at the level of 10^{-4}.

One way to see why frequency doubling occurs with β-waves is by considering the propagation matrix from a location just downstream of the error to a location s_1 where

$$\sin(\psi_1 - \psi_0) = 1 \tag{8.33}$$

The top right element in the propagation matrix M_{10} in Equation 3.31 becomes

$$m_{21} = \sqrt{\tilde{\beta}_1 \tilde{\beta}_0} \tag{8.34}$$

Since $\tilde{\beta}_0 = \beta_0$ is unperturbed by the thin quadrupole error, then β_1 must also be unperturbed, because m_{21} is unchanged, like all of the matrix elements of M_{10}. The frequency doubling in Equation 8.32 guarantees this.

In general, the horizontal β-function obeys the differential equation

$$\frac{d^2 b_x}{ds^s} + K\,b_x - b_x^{-3} = 0 \tag{8.35}$$

where

$$b_x = \sqrt{\beta_x} \tag{8.36}$$

and a similar equation (with negative K) applies for vertical β-function propagation. In a drift with $K = 0$

$$\beta(s) = \beta^* + \frac{s^2}{\beta^*} \tag{8.37}$$

in the case when β passes through a minimum at $s = 0$, for example at the centre of an experiment or an insertion device (see Exercise 3.5).

The β-wave that respects periodic boundary conditions in a circular accelerator due to a single small quadrupole strength perturbation

$$\frac{\Delta\beta}{\beta} = \frac{-\Delta q \beta_0}{2\sin(2\pi Q)} \cos(2|\psi - \psi_0| - 2\pi Q) \tag{8.38}$$

is similar to the closed orbit wave described by Equation 8.14. Figure 8.3 illustrates the β-wave cusp, and the double-speed phase advance.

The optics are especially sensitive to small quadrupole strength errors when the resonance denominator in Equation 8.38 approaches zero in either plane, when

$$p \approx 2Q_x \quad \text{or} \quad p \approx 2Q_y \tag{8.39}$$

That is, circular accelerator optics are vulnerable to half-integer resonances with $(q,r) = (2,0)$ or $(0,2)$.

Exercises

8.1 Consider a unit square in the tune plane (Q_x, Q_y) with corners at (n,n), $(n+1,n)$, $(n,n+1)$ and $(n+1,n+1)$.

a) On graph paper or with a computer program, draw the lines representing all sum resonances $p = q\,Q_x + r\,Q_y$ through fourth order – for positive integer values of q and r, with $q + r \leq 4$.
b) Plot all difference resonances $p = q\,Q_x - r\,Q_y$ through fourth order.
c) Where are the largest areas of tune space that are resonance-free?

8.2 Prove Equation 8.21, and extend it to second order in α.

8.3 The trigonometric law of sines states that

$$\frac{a}{\sin A} = \frac{b}{\sin B} = \frac{c}{\sin C} \tag{8.40}$$

where A, B, and C are the angles of a triangle, while a, b, and c are the lengths of the opposing sides.

a) Use the law of sines to show that Equation 8.17 guarantees the localisation of a three-bump.
b) What are the ratios of corrector strengths that close the three-bump if the phase advance between neighbouring correctors is 60 degrees, or 90 degrees?
c) What phase advance conditions make three-bump localisation difficult in practice? Why?

8.4 A quadrupole has strength $q = 1/f$ at a location with β-function β_0. What are the maximum relative quadrupole errors $\Delta q/q$ that are possible, as a function of unperturbed tune Q, before the accelerator becomes linearly unstable?

8.5 A β-three-bump can be constructed, by analogy to the closed orbit three-bump.

a) How can the strengths of three neighbouring quadrupoles be independently adjusted to change the β-function within them, but not outside (to first order in their strengths)?

b) How much does the tune Q change as a function of the strength of a β-three-bump?
c) What happens to the β-function in the other plane?
d) Can the β-function changes in both planes be simultaneously localised?

8.6 The interaction region quadrupole Q2 in RHIC has a focal length of about 3.0 m, at a location where the β-function is about 1400 m in collision optics with $\beta^* = 1$ m.

a) How accurately must the strength of this magnet be known and set, if the strength error must be guaranteed to generate a β-wave amplitude of less than 1%?
b) What tune shift is generated at this level of error?

8.7 Show by direct substitution that the general solution of horizontal motion

$$x = a\, b_x(s) \cos(\psi_x - \psi_0) \tag{8.41}$$

(where a and ψ_0 are constants) satisfies Hill's equation 4.1 if Equations 3.39 and 8.35 are valid.

9

Sextupoles, Chromaticity and the Hénon Map

Sextupoles are necessary in all but the smallest circular accelerators, in order to decrease the range of horizontal and vertical tunes that are spanned by particles with different momenta. Only if the tune spreads are small enough can all important resonances be avoided. But it is often the sextupoles themselves that are the strongest resonance drivers. This is the 'Catch-22' – sextupoles are necessary to avoid the resonances that are (often) most strongly driven by those same sextupoles.

The betatron motion of a particle that is off-momentum by $\delta = \Delta p/p$ in a purely linear lattice is described by

$$x'' + \frac{K}{(1+\delta)} x = 0 \qquad (9.1)$$

$$y'' - \frac{K}{(1+\delta)} y = 0$$

as previously discussed in Section 4.2. The local quadrupole strength $K(s)$ is (in effect) weaker for a higher momentum particle, and the betatron tunes naturally tend to decrease. With or without correction, the rate of variation of tune with momentum is measured by the chromaticity χ, where

$$\chi \equiv \frac{dQ}{d\delta} \qquad (9.2)$$

Natural chromaticity, chromaticity correction and the inevitable resonance driving side effects are conveniently demonstrated by a pure FODO lattice.

9.1 Chromaticity in a FODO Lattice

The tune of an accelerator with a phase advance of ϕ in each of N FODO cells is

$$Q = \frac{N}{2\pi} \phi \qquad (9.3)$$

so that the natural (uncorrected) chromaticity is

$$\chi_{NAT} = \frac{N}{2\pi} \frac{d\phi}{d\delta} \tag{9.4}$$

If the FODO quadrupoles D and F have equal and opposite strengths, then the phase advance per cell is the same in both planes, and

$$\frac{d\phi}{d\delta} = -2 \tan(\phi/2) \tag{9.5}$$

according to Equation 3.48, so that the natural horizontal and vertical chromaticities are

$$\chi_{x,NAT} = \chi_{y,NAT} = -Q \frac{\tan(\phi/2)}{\phi/2} \tag{9.6}$$

This illustrates a convenient rule of thumb: natural chromaticities are about the same size as (or rather larger than) the design tunes, with a negative sign. This rule is least accurate in storage ring colliders, where the β-squeeze optics sections near to the collision points also contribute strongly, perhaps doubling the natural chromaticities coming from the arcs.

If (using typical RHIC values) the natural chromaticity is about -50, and the root mean square relative momentum spread σ_p/p is about 2×10^{-3}, then the uncorrected chromatic tune spread is about

$$\sigma_Q = |\chi| \frac{\sigma_p}{p} = 0.1 \tag{9.7}$$

This is an unacceptably large footprint in the (Q_x, Q_y) tune plane. Fortunately, it is easy to reduce the chromaticities and the tune footprint by adding sextupoles immediately next to quadrupoles.

The total horizontal deflection due to a thin quadrupole–sextupole pair is illustrated in Figure 9.1. The closed orbit is horizontally displaced by $\eta\delta$, where η is the dispersion, so that the total horizontal displacement is

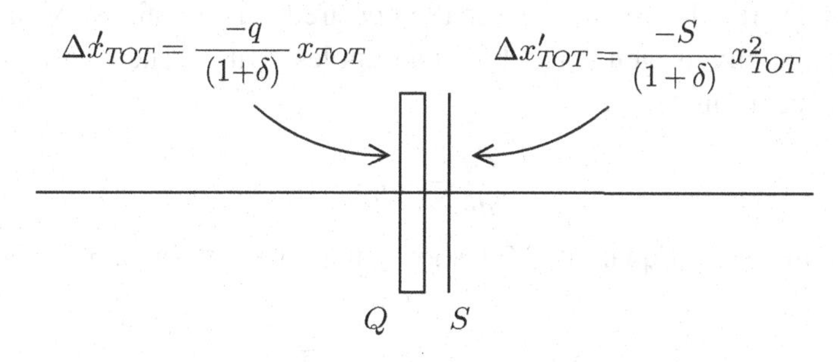

Figure 9.1 A thin chromaticity correction sextupole of integrated strength S placed immediately next to a quadrupole of strength q.

$$x_{TOT} = \eta\,\delta \, + \, x \tag{9.8}$$

and to first order in δ the net horizontal deflection is

$$\Delta x'_{TOT} = -q(1-\delta)(\eta\delta + x) \; - \; S(1-\delta)(\eta\delta + x)^2 \; + \; \mathcal{O}(\delta^2) \tag{9.9}$$

where q and S are the integrated quadrupole and sextupole strengths. Separating out the different components of order $x^n\delta^m$ gives

$$\begin{aligned}(\Delta\eta')\delta + \Delta x' = & -q.\eta\delta \\ & -q.x \; - \; S.x^2 \\ & +q.x\delta - S.2\eta.x\delta + S.x^2\delta \\ & +\mathcal{O}(\delta^2)\end{aligned} \tag{9.10}$$

where the first row (of order $x^0\delta^1$) describes how the dispersion slope η' changes, while the second row (of order $x^m\delta^0$) contains the so-called geometric terms.

Extracting all the terms of order $x^1\delta^m$ shows that the net linear quadrupole effect is

$$\Delta x' = -q.x + (q - 2\eta S)x\delta \tag{9.11}$$

showing that the quadrupole strength is constant with respect to momentum – and so the net chromaticities are zero – simply if

$$S = \frac{q}{2\eta} \tag{9.12}$$

at every D and F quadrupole–sextupole pair. Typically the F dispersion value in a FODO cell is twice the D value, as illustrated in Figure 4.3, so that the chromaticity sextupole strength S_F is typically half the strength (and the opposite sign) of S_D.

9.2 Chromaticity Correction

At least two families of chromaticity sextupoles are needed in order to set both horizontal and vertical chromaticities to their desired (small) values. For any optics, natural chromaticity is generated by quadrupoles that weaken with increasing momentum, according to

$$\frac{dq}{d\delta} = -q \tag{9.13}$$

Combining this with Equation 8.29 shows that the natural chromaticities are given by

$$\begin{pmatrix} \chi_x \\ \chi_y \end{pmatrix}_{NAT} = -\frac{1}{4\pi}\begin{pmatrix} \sum q\beta_x \\ \sum -q\beta_y \end{pmatrix} \tag{9.14}$$

where the sums (over all quadrupoles) are readily converted to integrals if the quadrupoles are thick. The quadrupole strength q is usually positive at F locations with large β_x values, and negative when β_y is large at D locations, so that both natural chromaticities are negative.

Similarly, sextupoles contribute a quadrupole component

$$\frac{dq}{d\delta} = 2\eta S \tag{9.15}$$

so that the chromaticity correction due a single family of many sextupoles of strength S_1 is

$$\Delta\begin{pmatrix} \chi_x \\ \chi_y \end{pmatrix} = \frac{S_1}{2\pi}\begin{pmatrix} \sum \eta\beta_x \\ \sum -\eta\beta_y \end{pmatrix} \tag{9.16}$$

A sextupole family that increases the horizontal chromaticity necessarily decreases the vertical chromaticity, and vice versa, because of the sign change between the two terms. Fortunately it is possible to place one family at F locations (where η and β_x are larger, and β_y is smaller), and a second family at D locations (where the relative roles are swapped).

With two families of sextupoles, the chromaticities are set to goal values by solving the equation

$$\begin{pmatrix} \chi_x \\ \chi_y \end{pmatrix}_{GOAL} = 2\begin{pmatrix} \sum_F \eta\beta_x & \sum_D \eta\beta_x \\ \sum_F -\eta\beta_y & \sum_D -\eta\beta_y \end{pmatrix}\begin{pmatrix} S_F \\ S_D \end{pmatrix} + \begin{pmatrix} \chi_x \\ \chi_y \end{pmatrix}_{NAT} \tag{9.17}$$

for S_F and S_D. Clearly there is no advantage in placing chromaticity correction sextupoles at locations where the dispersion η is small – for example, in injection and extraction straights. Fortunately it is usually easy to place sextupoles next to quadrupoles in the arcs, where the dispersion η is relatively large. In practice, the goal chromaticities are slightly positive, perhaps with $\chi \approx 2$, in order to combat intensity-independent instabilities like the head–tail effect [5].

In RHIC, correcting the chromaticities in this way reduces the chromatic root mean square tune spread σ_Q from about 0.1 to about 0.004. Note that if the tunes vary only with momentum, then the tune spread lies on a line

$$(Q_x(\delta), Q_y(\delta)) = (Q_x(0), Q_y(0)) + (\chi_x, \chi_y)\,\delta \tag{9.18}$$

that is straight to first order in δ. In practice, the two tunes both also vary with horizontal and vertical betatron amplitudes, so that the 1-D line broadens into a 2-D tune spread footprint.

Reducing the total tune spread from all sources to less than about ± 0.01 makes dramatically more working points (Q_x, Q_y) available in the tune plane, avoiding sum resonances up to fifth order, as illustrated in Figure 9.2. This begs some obvious

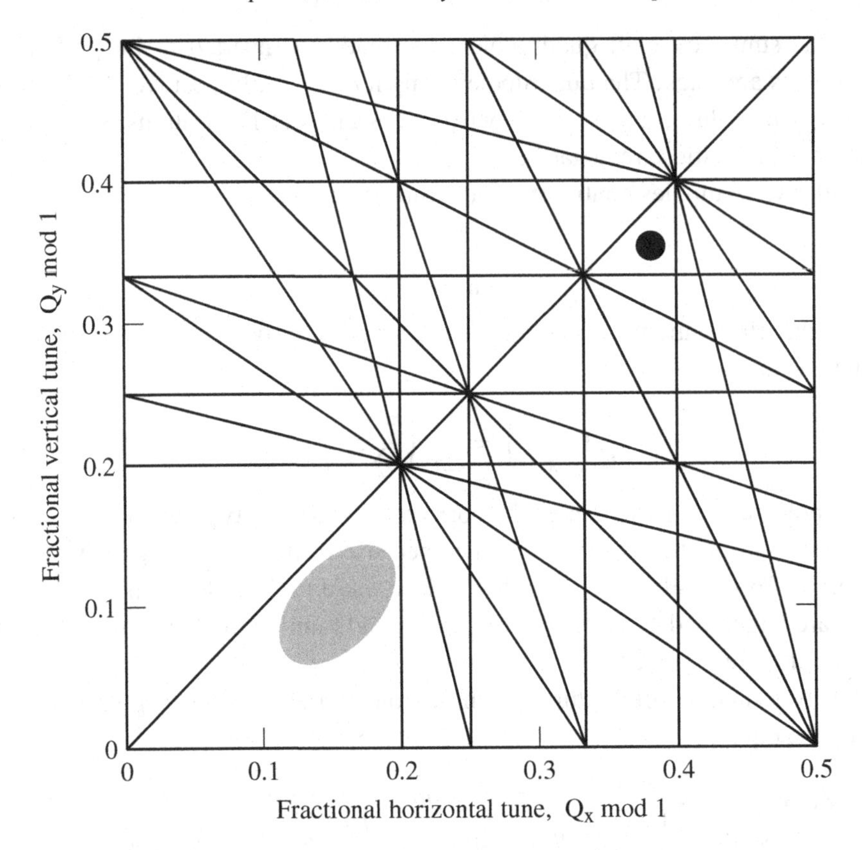

Figure 9.2 Tune plane resonance lines $p = qQ_x + rQ_y$ with q and r positive, up to resonance order $q + r = 5$. The grey footprint with a tune spread of ± 0.05 near $(Q_x, Q_y) = (0.15, 0.10)$ is due, for example, to uncorrected natural chromaticities. It can only avoid straddling the (unshown) sixth order resonances by moving the working point dangerously close to an integer resonance, $p = Q_y$ or $p = Q_x$. The black footprint near $(Q_x, Q_y) = (0.37, 0.35)$, with a tune spread of about ± 0.005, can be placed in many more locations without touching potentially dangerous resonance lines.

questions: What drives the resonance lines? How strong are they? Is it sufficient to avoid only resonances up to fifth order? Do sextupoles drive only third-order resonances (as first-order perturbation theory predicts)? How do sextupoles cause the tunes to deviate as a function of horizontal and vertical betatron amplitudes?

9.3 The Hénon Map – A Unit Strength Sextupole in 1-D

Michel Hénon, a mathematician and astronomer, was one of the first to investigate nonlinear maps numerically. In 1969, he wrote that the map that now bears his name:

exhibits all the typical properties of more complicated mappings and dynamical systems [19]

until finished { (9.19)

$$\begin{pmatrix} z \\ z' \end{pmatrix} = \begin{pmatrix} \cos(2\pi Q) & \sin(2\pi Q) \\ -\sin(2\pi Q) & \cos(2\pi Q) \end{pmatrix} \begin{pmatrix} z \\ z' \end{pmatrix} + \begin{pmatrix} 0 \\ z^2 \end{pmatrix}$$

}

The only control parameter in this map, Q, is analogous to the horizontal tune of an accelerator that contains a single sextupole of unit strength.

One-turn motion in normalised phase space (x, x') at a reference point just before a sextupole of strength g is a kick

$$\begin{pmatrix} x \\ x' \end{pmatrix}_{0+\epsilon} = \begin{pmatrix} x \\ x' - gx^2 \end{pmatrix}_0 \tag{9.20}$$

followed by a rotation described by the matrix R

$$\begin{pmatrix} x \\ x' \end{pmatrix}_1 = R(2\pi Q) \begin{pmatrix} x \\ x' \end{pmatrix}_{0+\epsilon} \tag{9.21}$$

If the tune is close to 1/3 (ignoring the integer component), so that

$$Q = \frac{1}{3} + \delta Q \tag{9.22}$$

then the small net motion after three turns is given to first order in g by

$$\begin{pmatrix} x \\ x' \end{pmatrix}_3 - \begin{pmatrix} x \\ x' \end{pmatrix}_0 \approx \tag{9.23}$$

$$3\mu \begin{pmatrix} x' \\ -x \end{pmatrix}_0 - g\left[\begin{pmatrix} S_3 \\ C_3 \end{pmatrix} x_0^2 + \begin{pmatrix} S_2 \\ C_2 \end{pmatrix} x_1^2 + \begin{pmatrix} S_1 \\ C_1 \end{pmatrix} x_2^2\right]$$

where

$$\mu = 2\pi\,\delta Q, \quad C_k = \cos\left(k\frac{2\pi}{3}\right), \quad S_k = \sin\left(k\frac{2\pi}{3}\right) \tag{9.24}$$

This is written more succinctly as a three-turn discrete Kobayashi Hamiltonian

$$H_3 = \frac{\mu}{2}(x^2 + x'^2) + \frac{g}{9}\sum_{k=1}^{3}(C_k x + S_k x')^3 \tag{9.25}$$

where the equations of three-turn motion are

$$\Delta x \approx \frac{\partial H_3}{\partial x'} \Delta t \qquad (9.26)$$
$$\Delta x' \approx -\frac{\partial H_3}{\partial x} \Delta t$$

with $\Delta t = 3$ [25, 40, 43].

The value of H_3 is approximately constant for each trajectory when the net three-turn motion is small: when μ and the amplitude are small enough. Factorising Equation 9.25

$$H_3 = \frac{g}{3}\left(\frac{2\mu}{g}\right)^3 + \frac{g}{12}\left(x + \sqrt{3}x' + \frac{4\mu}{g}\right)\left(x - \sqrt{3}x' + \frac{4\mu}{g}\right)\left(x - \frac{2\mu}{g}\right) \qquad (9.27)$$

then leads to the prediction that trajectories with a constant value

$$H_3 = \frac{g}{3}\left(\frac{2\mu}{g}\right)^3 \qquad (9.28)$$

lie on straight lines that intersect to form an equilateral triangle in normalised phase space. Figure 9.3a confirms this prediction: circles at very small amplitudes morph into triangles at moderate amplitudes. In contrast, the appearance of three isolated islands at large amplitudes is entirely unexpected in the discrete Hamiltonian framework, to first order in g.

9.4 A Taxonomy of 1-D Motion

The Hénon map exhibits the taxonomy of four different kinds of phase-space trajectories that are observed in 1-D simulations, in general. These behaviours are variously avoided, or exploited, in accelerators. Figure 9.3 illustrates this taxonomy, following the motion of many trajectories launched with different initial amplitudes and phases for four different values of the control parameter Q, each value close to a different rational fraction $1/N$.

Regular Non-Resonant Trajectories

Motion is regular, and the turn-by-turn trajectories fall on roughly circular continuous lines, at small amplitudes near the origin of each of the panels in Figure 9.3. A particle launched in this region comes arbitrarily close to any given phase angle around the origin, given enough time (iterations, or turns). The beam fills a region around the origin with an area proportional to its emittance, when an accelerator is operating normally with Q well away from all low-order rational fractions.

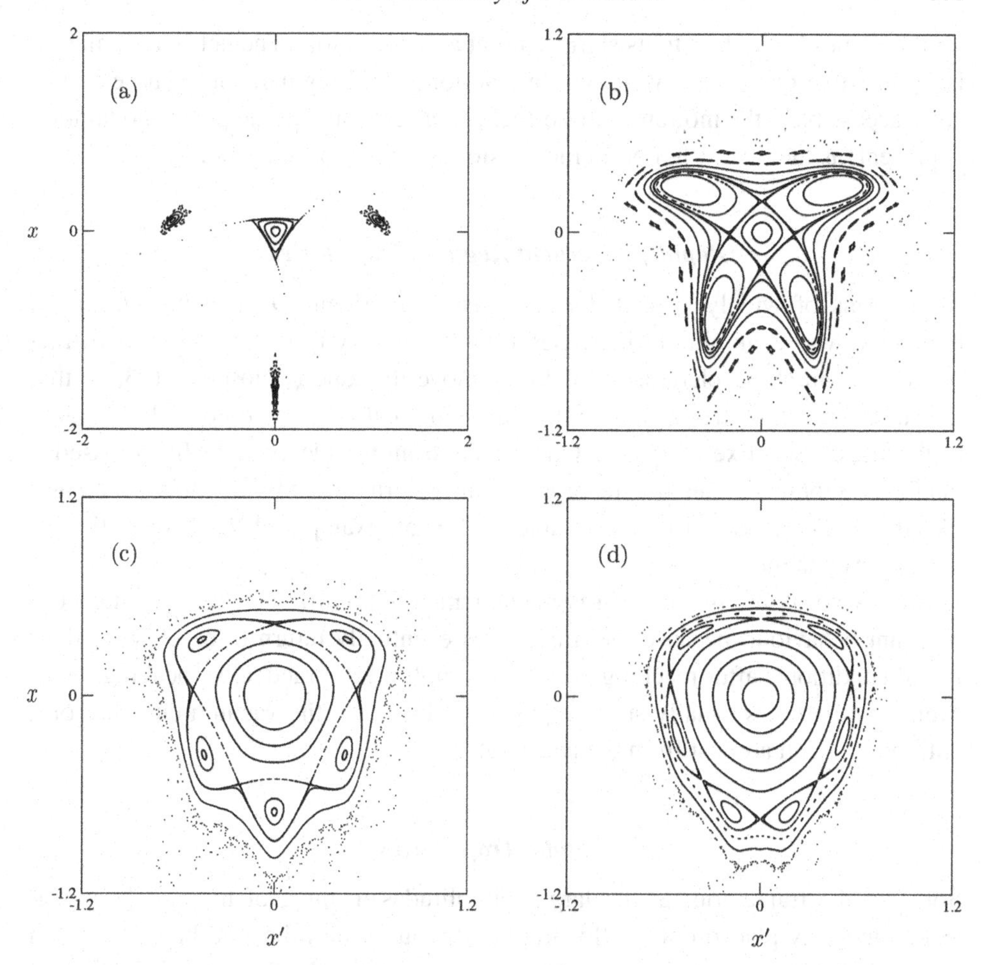

Figure 9.3 Turn-by-turn behaviour under the Hénon map in normalised phase space, for four different values of Q that are close to low order rational fractions. The sole control parameter Q is analogous to the horizontal tune in a lattice with a single sextupole. When the tune is $Q = 0.324 \approx 1/3$ in (a), the equilateral triangle and the motion within it (but not the three outlying islands) are well described by first order discrete Hamiltonian theory. Four large islands appear in (b) when $Q = 0.2516 \approx 1/4$. A trajectory launched in one of the five resonance islands in (c), when $Q = 0.211 \approx 1/5$, hops from one island to the next, until the net motion is small after five turns. Six islands are almost rotationally symmetric in (d) when $Q = 0.185 \approx 1/6$. The six islands resemble each other and the standard map structure shown in Figure 4.6, when plotted in action-angle space.

Regular Resonant Trajectories

The non-circular distortions increase as the amplitude increases, until the motion breaks up into a chain of resonances islands. For example, a trajectory launched in the middle of one of the five islands in Figure 9.3c will hop from island to island, until it comes back close to its starting point after five turns. The net N-turn motion is small when $Q \approx I/N$. Although this motion is still regular, some phase angles are inaccessible – the motion is phase-locked, or resonant. Some phase modulation is present, as the five-turn motion moves slowly around each island.

Rapidly Divergent Regular Trajectories

Three arms of widely spaced dots are visible in Figure 9.3a, demonstrating a trajectory whose amplitude increases rapidly, turn-by-turn. One way to extract particles from an accelerator is to slowly move the tune Q closer to $1/3$, so that the stable area near the origin shrinks, squeezing particles out onto such divergent trajectories. (See Exercise 9.4.) Once a large amplitude particle has crossed a magnetic septum it can be steered into an external beamline. Slow extraction like this is discussed further in Chapter 10, as an example of the exploitation of nonlinear dynamics.

Strictly speaking, this motion is mathematically no different from regular non-resonant behaviour, because the trajectories eventually return to be close to their initial conditions, after reaching very large amplitudes. Practically speaking, however, such particles usually strike a physical obstacle – the vacuum chamber or a collimator – and are lost from the accelerator.

Chaotic Trajectories

Some of the trajectories at the largest amplitudes in three of the panels appear to be randomly placed dots, following no obvious pattern beyond being confined to certain regions of phase space. Chaotic trajectories like these are immediately recognisable. It is tempting to agree with Stewart:

I shall not today attempt further to define … that shorthand description, and perhaps I could never succeed in intelligibly doing so. But I know [chaos] when I see it, … [51].

except that a semi-rigorous definition of chaos is easy. A pair of trajectories launched extremely close together in a regular region of phase space diverges linearly, when time t becomes large enough. In contrast, a pair launched in a chaotic region diverges exponentially, so that

$$D(t) \sim D_0 \exp(\lambda t) \tag{9.29}$$

where the phase space distance $D(t)$ separating the pair is given by

$$D^2 = (x_A - x_B)^2 + \left(x'_A - x'_B\right)^2 \tag{9.30}$$

and λ is the Lyapunov exponent.

9.5 Dynamic Aperture

A particle that strikes a physical obstacle in an accelerator – usually the beam pipe – is lost forever. Often it is not necessary to simulate these physical apertures very accurately, because doomed particles follow rapidly divergent trajectories, and are quickly lost whether the aperture is 0.1 m or 1.0 m. In this case the dynamic aperture found by simulation is significantly less than – and independent of – the physical aperture.

The horizontal dynamic aperture of a lattice with a single sextupole (the Hénon map) depends strongly on the tune, as shown in Figure 9.4 over the range $0 < Q < 0.5$. Clearly the maximum initial amplitude for stable motion shrinks whenever the tune approaches a low-order rational fraction. The severity of these 1-D resonances declines as the denominator increases: $Q \approx 1/3$ has a much more dramatic effect than (say) $Q \approx 2/7$. This is fortunate, because the number line is dense in rational fractions, and the 2-D tune plane is dense in resonance lines.

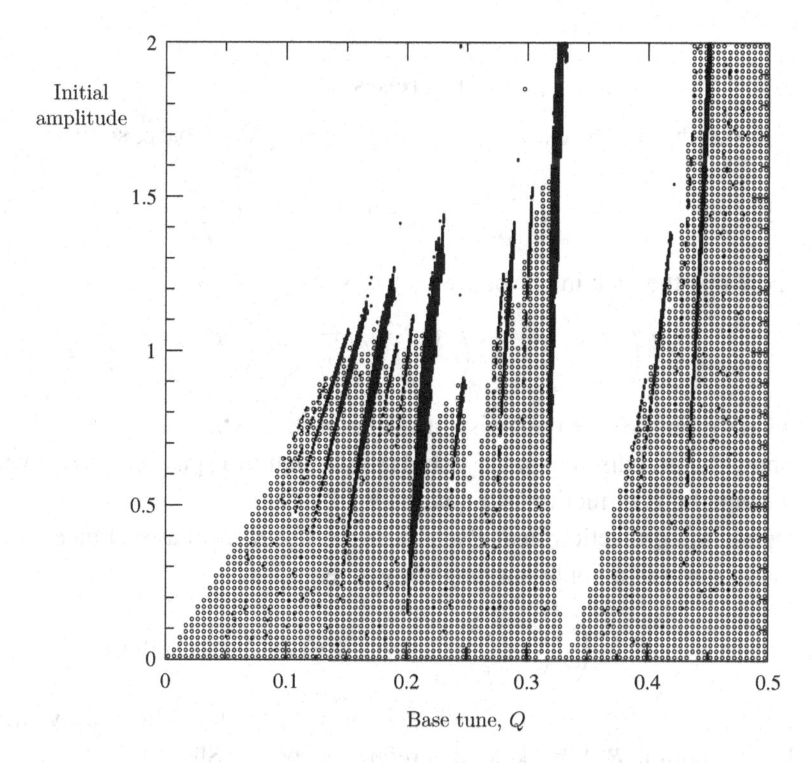

Figure 9.4 Dynamic aperture versus tune under the Hénon map. Even a single sextupole drives all resonances $Q \approx I/N$. The area that is lightly shaded indicates initial conditions that lead to stable non-resonant motion, while the darker area indicates resonant island motion up to order $N = 20$. Trajectories launched in the unshaded area are quickly lost forever.

Chapter 8 discusses how the sensitivity to dipole and quadrupole strength errors becomes extreme for some values of Q, concluding that dipole and quadrupole errors drive first- and second-order resonances. The first-order nonlinear theory presented in this chapter correctly predicts how one sextupole (or many) drives $Q \approx 1/3$, with success at moderate amplitudes, but failing at large amplitudes. Worse, first-order theory fails completely to predict the resonances that generate island chains and reduce the dynamic aperture, at tunes that are not close to 1/3.

Higher-order perturbation theory has some success in extending the reach of analytical prediction after a significant effort, especially in one dimension [49, 33]. And first-order perturbation theory can be extended to include 2-D behaviour, predicting the strength of low-order resonance lines in the tune plane. But in the end it is hard to disagree with Taff:

Beyond first-order theory I know of no useful result from perturbation theory in celestial mechanics ..... Frequently the second approximation produces nonsensical results... [52].

Exercises

9.1 If the strength of a sextupole is g_p in physical phase space, so that

$$\begin{pmatrix} \Delta x'_p \\ \Delta y'_p \end{pmatrix} = -g_p \begin{pmatrix} x_p^2 - y_p^2 \\ -2x_p y_p \end{pmatrix} \tag{9.31}$$

show that the kick in normalised phase space is

$$\begin{pmatrix} \Delta x' \\ \Delta y' \end{pmatrix} = -\begin{pmatrix} (\beta_x^{3/2} g_p)\, x^2 - (\beta_x^{1/2}\beta_y g_p)\, y^2 \\ (\beta_x^{1/2}\beta_y g_p)\, xy \end{pmatrix} \tag{9.32}$$

where β_x and β_y are the Twiss functions at the sextupole.

9.2 Show that the discrete Kobayashi Hamiltonian of Equation 9.25 generates the equations of motion of Equation 9.23.

9.3 Define the projection map P as linear motion R from a reference point to the location of a nonlinear magnet, where

$$R = \begin{pmatrix} c_x & s_x \\ -s_x & c_x \end{pmatrix} \qquad c_x = \cos(\phi_x) \quad s = \sin(\phi_x)$$

followed by the nonlinear kick $\Delta x' = gx^n$, finally followed by inverse linear motion R^{-1} back to the reference point. Show that the discrete projection Hamiltonian representing P is given by

$$H_p = -\frac{g}{n+1}(c_x x + s_x x')^{n+1} \tag{9.33}$$

9.4 Consider the equilateral triangle in (x, x') normalised phase space predicted by Equations 9.27 and 9.28.

a) What is the radius of the largest circle that can be inscribed inside the triangle?
b) What is the orientation of the triangle?
c) What happens to the area and the orientation of the triangle as the tune Q is (slowly) swept through the value of 1/3?

9.5 Investigate motion under the Hénon map by writing code that permits you to adjust the tune Q, and allows you to launch trajectories at any initial location in phase space. Consider the plot of Hénon dynamic aperture (DA) versus tune shown in Figure 9.4.

a) How small must $|Q - 1/3|$ be, for the predictions of Exercise 9.4 to be reasonably valid?
b) Devise and define a convenient quantitative measure of the size of the stable region – the dynamic aperture, DA. (There are many ways to do this.)
c) How does the DA vary as a function of the time tracked, for interesting sample tunes in the range $0 < Q < 0.5$?
d) How does the DA vary as a function of simulated physical aperture?
e) How does the DA in the range $0.5 < Q < 1.0$ relate to the DA in the range below $Q = 0.5$? Why?

9.6 Consider an accelerator made entirely of FODO cells of length $2L$ and bend radius R, with the same phase advance per cell $\Delta\phi$ in both planes. If the identical cells are all perfect, then it is only necessary to simulate a single cell with two sextupoles, in a natural extension of the Hénon map.

a) If both net chromaticities are set to zero according to Equation 9.12, how does the horizontal DA vary with $\Delta\phi$?
b) How does the DA vary as a function of L and R (with zero net chromaticities), for a fixed value of $\Delta\phi$?
c) Suppose that both sextupoles are strengthened by the same scale factor κ (at fixed L, R, and $\Delta\phi$). Compare simulation and theory to demonstrate and explain how the DA varies with κ. Hint: consider the normalised phase space co-ordinate transformation

$$\begin{pmatrix} X \\ X' \end{pmatrix} = \kappa^m \begin{pmatrix} x \\ x' \end{pmatrix} \tag{9.34}$$

d) Set the FODO cell tune $\Delta\phi/2\pi$ near a low-order rational fraction in a 2-sextupole simulation, so that a chain of resonance islands appears, centred at a moderate phase space amplitude. How does this amplitude vary with κ?

9.7 (See also Exercise 4.6.) Investigate motion under the Hénon map by following and recording the motion of two trajectories launched very close together, at any initial location in phase space.

a) Measure the Lyapunov exponent λ for a set of different initial locations, moving between areas of regular and chaotic motion.
b) Measure the Lyapunov exponent λ at a fixed initial location, for a set of tunes that moves the dynamics from regular to chaotic.
c) Approximately how many turns must be tracked, in order to distinguish between regular and chaotic motion?

9.8 The map of a single octupole with strength G

$$\begin{pmatrix} x \\ x' \end{pmatrix} = \begin{pmatrix} c & s \\ -s & c \end{pmatrix} \begin{pmatrix} x \\ x' \end{pmatrix} \qquad c \equiv \cos(2\pi Q) \quad s \equiv \sin(2\pi Q)$$
$$x' = x' - G\,x^3 \tag{9.35}$$

is closely related to the single sextupole Hénon map.

a) Find the normalised phase space co-ordinate transformation that leaves the single octupole map independent of G.
b) Predict how the resonance island centre amplitude scales as a function of G.
c) Confirm or disprove this scaling with a numerical experiment.
d) What is the relationship between phase space behaviour just below and just above $Q = 0.5$?

10
Octupoles, Detuning and Slow Extraction

Most sources of nonlinear motion – like chromaticity correction sextupoles, magnetic field imperfections and the beam–beam interaction – are unavoidable and unfortunate. On the other hand, octupoles (or sextupoles) can be deliberately added to enable the controlled slow spill of particles, perhaps to a high-energy physics beamline, or to a patient in a treatment room. Slow extraction with octupoles demonstrates the practical use of first-order perturbation theory to design nonlinear optics.

10.1 Single Octupole Lattice

One-turn motion in the single octupole lattice shown in Figure 10.1 bears a close analogy to the Hénon (single sextupole) motion discussed in Section 9.3. Starting at a reference point just before the octupole, there is first an angular kick that is given in physical space by

$$\Delta x_p' = -g_p x_p^3 \tag{10.1}$$

and in normalised phase space (without a subscript) by

$$\Delta x' = -g x^3 \tag{10.2}$$

where the strength g depends on the horizontal β-function according to

$$g = g_p \beta^2 \tag{10.3}$$

This kick causes an additional phase advance of

$$\begin{aligned} \Delta\phi &= g a^2 \sin^4(\phi) \\ &= g a^2 \left(\frac{3}{8} - \frac{1}{2}\cos(2\phi) + \frac{1}{8}\cos(4\phi) \right) \end{aligned} \tag{10.4}$$

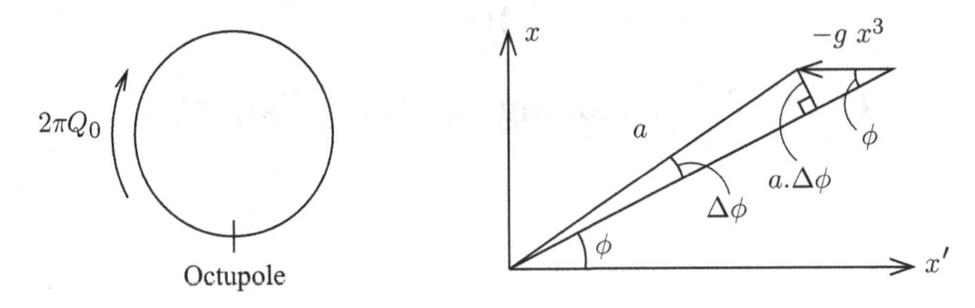

Figure 10.1 A single octupole lattice, and the effect of a single octupole in normalised phase space.

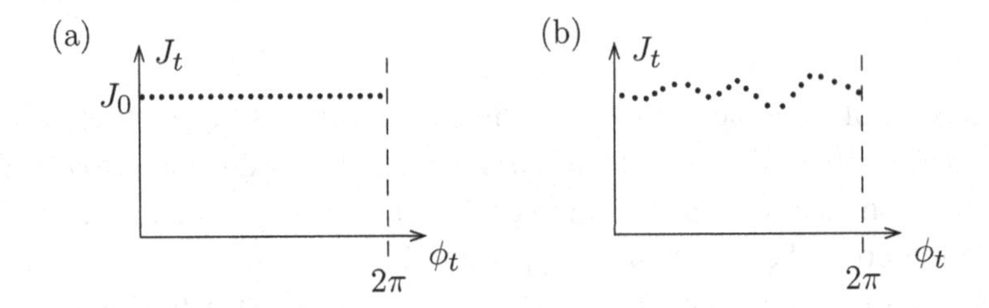

Figure 10.2 The linear motion in 1-D action-angle space shown in (a) reaches arbitrarily close to any phase ϕ given sufficient time, with a smooth and even distribution of phases at constant action J_0, if the tune Q_0 is not a rational fraction. When nonlinearities are introduced the motion generally falls on a KAM surface $J_t = J(J_0, \phi_t)$, as shown in (b), if the motion is neither resonant (Q_0 is not a rational fraction) nor chaotic.

according to the normalised phase space geometry shown in Figure 10.1. One-turn motion is completed by following the thin octupole kick with a phase space rotation of $2\pi Q_0$.

Except for the effect of the octupole, the amplitude a is constant and the turn-by-turn phase

$$\phi_t = \phi_0 + 2\pi Q_0 t \tag{10.5}$$

is smooth and evenly distributed, given enough time (and if Q_0 is not a rational fraction). This is illustrated in Figure 10.2. In the approximation that ϕ is evenly distributed even with the octupole turned on, the average additional phase advance per turn as $t \to \infty$ is

$$\langle \Delta\phi \rangle = ga^2 \left\langle \sin^4(\phi) \right\rangle = \frac{3}{8}\, ga^2 \tag{10.6}$$

so that the tune changes with average amplitude according to

$$Q = Q_0 + \frac{3}{16\pi}\, ga^2 \tag{10.7}$$

Such detuning with amplitude is absent, to first order in g, for all multipoles (like sextupoles and decapoles) that have an even power $2m$ in the exponent of the kick, since

$$\langle \sin^{2m+1}(\phi) \rangle = 0 \tag{10.8}$$

This makes it even more remarkable that the discrete Kobayashi Hamiltonian formalism in Section 9.3 correctly predicts the triangular phase space distortions for tunes both above and below $Q = 1/3$, in the presence of sextupoles. (See Exercise 9.4.)

10.2 Discrete Motion in Action-Angle Space, (J, ϕ)

Detuning in the action-angle representation becomes

$$Q = Q_0 + \frac{3}{8\pi} gJ \tag{10.9}$$

Detuning itself can lead to filamentation and emittance blow-up, through shear motion around normalised phase space circles that is described, for example, in Section 5.3. But in addition to detuning, nonlinear elements like octupoles also distort the turn-by-turn (J, ϕ) trajectories from the flatline on the left of Figure 10.2, into the wavy line on the right. If the perturbed nonlinear 1-D motion is non-resonant, is not rapidly divergent, and is not chaotic, then for a general trajectory labelled by J_0 it is

$$J_t = J_0 + \sum_{k=1}^{\infty} u_k \cos(k\phi_t + \phi_k) \tag{10.10}$$

$$\phi_t = \phi_0 + 2\pi Q t + \sum_{k=1}^{\infty} v_k \cos(k\, 2\pi Q t + \theta_k)$$

where u_k, v_k, ϕ_k and θ_k are constants (for each k), and where the tune is

$$Q \equiv \lim_{T\to\infty} \left(\frac{\phi_T - \phi_0}{2\pi T} \right) \tag{10.11}$$

If it exists, then such a monotonic contour $J_t = J(J_0, \phi_t)$ is properly called a Kolmogorov–Arnold–Moser (KAM) surface. Regions of action-angle space at larger and smaller actions are rigorously separated by KAM surfaces, even if the motion in those regions is resonant, divergent or chaotic [12]. This is illustrated in Figures 4.6 and 9.3.

It is often possible to find a one-turn difference Hamiltonian H_1 that succinctly reproduces one-turn motion through the canonical equations

$$\Delta\phi = \frac{\partial H_1}{\partial J} \tag{10.12}$$
$$\Delta J = -\frac{\partial H_1}{\partial \phi}$$

For example, the one-turn difference Hamiltonian

$$H_1 = 2\pi Q_0 J \; + \; \alpha J^2 \tag{10.13}$$

reproduces purely linear behaviour with an amplitude independent tune Q_0 when $\alpha = 0$, or the pure shear motion shown in Figure 5.3 when $\alpha = \pi\, dQ/dJ$, or the octupole detuning of Equation 10.9 when $\alpha = 3g/8$. Similarly, the generalised 2-D one-turn Hamiltonian

$$H_1 = 2\pi(Q_{x0}\, J_x \; + \; Q_{y0}\, J_y) \; + \; \sum_{ijkl} V_{ijkl}\, J_x^{i/2} J_y^{j/2} \sin(k\phi_x + l\phi_y + \phi_{ijkl}) \tag{10.14}$$

reproduces the distorted 1-D motion described in Equation 10.10 when V_{ijkl} is non-zero only for purely horizontal terms with $j = l = 0$ [40].

10.3 Two-Turn Motion with $Q \approx 1/2$

The canonical equations of one-turn motion in Equation 10.12 are closely analogous to those in Equation 9.26, for three-turn motion with $Q \approx 1/3$ in the presence of a single sextupole. However, the net one-turn motion is not small, and the quantity H_1 is far from being a constant of the motion. If the tune is close to 1/2 (ignoring the integer component), so that

$$Q = \frac{1}{2} \; + \; \delta Q \tag{10.15}$$

then matrix manipulation shows that the small net two-turn motion with a single octupole is represented in normalised phase space by the two-turn Kobayashi Hamiltonian

$$H_2 = 2\pi\; \delta Q \,\frac{1}{2}(x^2 + x'^2) \; + \; \frac{g}{4} x^4 \tag{10.16}$$

where g is the strength of the octupole. In action-angle phase space this becomes

$$H_2 = 2\pi\; \delta Q\, J \; + \; \left[\frac{3}{8} - \frac{1}{2}\cos(2\phi) + \frac{1}{8}\cos(4\phi)\right] gJ^2 \tag{10.17}$$

or more generally, in the presence of many octupoles

$$H_2 = 2\pi\; \delta Q\, J \; + \; [V_0 + V_2 \cos(2\phi + \phi_2) + V_4 \cos(4\phi + \phi_4)]\; J^2 \tag{10.18}$$

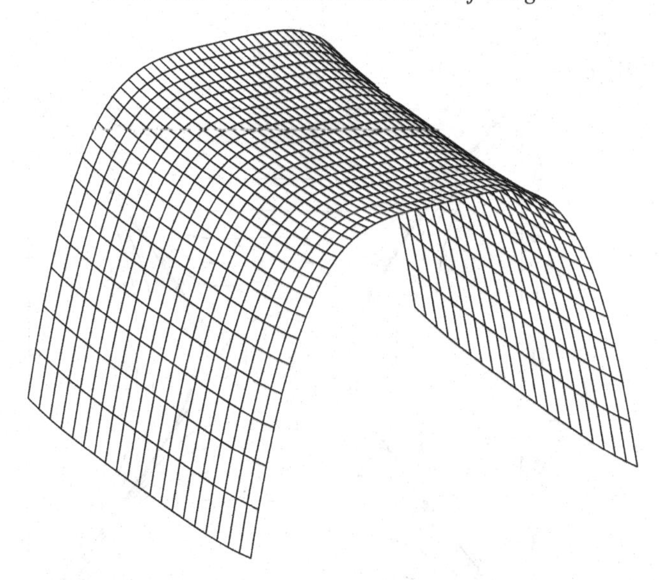

Figure 10.3 The profile of the two-turn Hamiltonian H_2 in normalised phase space (x, x') graphically describes how slow extraction near a half-integer tune $Q \approx 1/2$ is initiated, using octupoles. The value of H_2 is an approximate invariant of the motion for individual protons.

The constants V_0, V_2, V_4, ϕ_2 and ϕ_4 are controlled to first order by the strengths and locations of the octupoles around an accelerator. They are also influenced – to second order in strength – by chromaticity correction sextupoles.

The quantity H_2 is an approximate invariant of the motion, so that particles flow along its contours. The 3-D visualisation of H_2 (and its 2-D contours) directly aids the comprehension and design of the slow extraction of protons with octupoles, as illustrated by Figures 10.3 and 10.4.

10.4 Slow Extraction near the Half-Integer

Octupoles drive 120 GeV protons across the wires of an electrostatic septum in the first stage of slow extraction from the Fermilab Main Injector, as illustrated in Figure 10.5 [22, 54]. The second-stage strong magnetic septum shown in Figure 10.6 diverts protons into an extraction channel.

The shallow stable pool near the origin of $H_2(x, x')$ in Figure 10.3 stores beam that has not yet been extracted. Protons are forced to leak over the edge of the pool by slowly (and carefully) decreasing δQ, thereby decreasing both the curvature and the area of the pool. Leaving the pool, protons follow a rapidly divergent trajectory, accelerating along a steepening side wall contour with a speed

$$v_2 = \left| \vec{\nabla} H_2 \right| \tag{10.19}$$

that is equal to the slope of H_2, as discussed in Section 4.4.

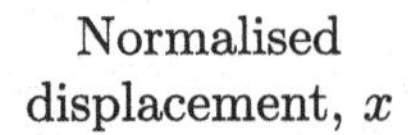

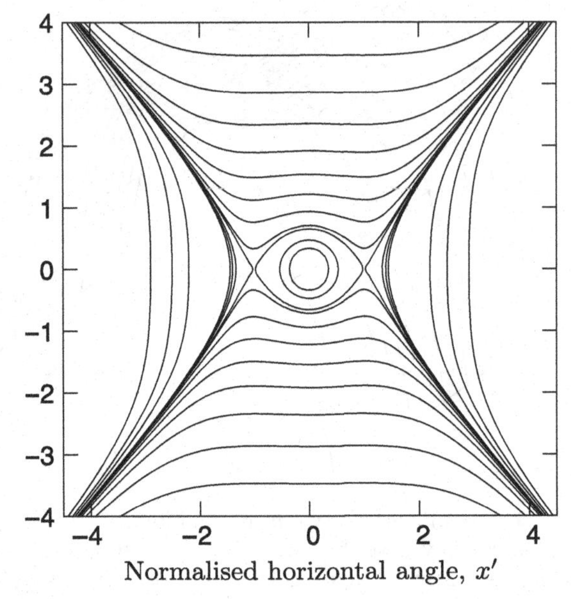

Figure 10.4 Contours of $H_2(x, x')$. Protons that are squeezed out of the stable separatrix near $(x, x') = (0, 0)$ stream along rapidly divergent trajectories that follow the narrow arms towards large amplitudes.

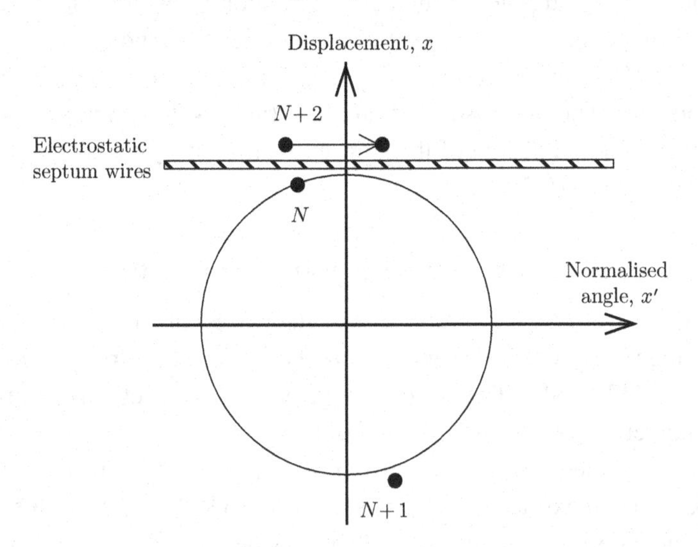

Figure 10.5 The phase space view of a proton on a rapidly divergent trajectory jumping across the wires of an electrostatic septum in two turns, during slow extraction.

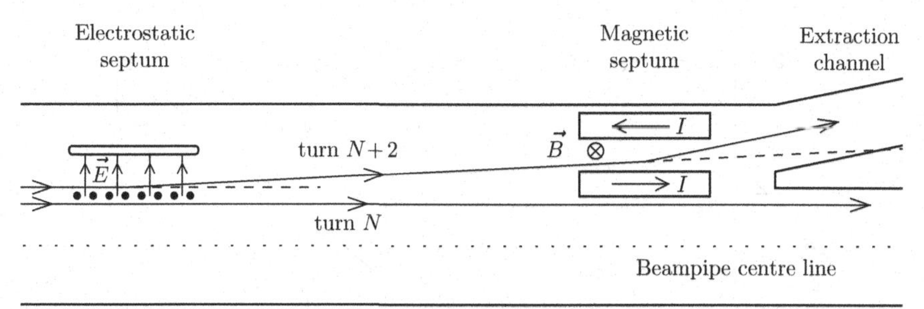

Figure 10.6 Plan view of a proton crossing the narrow wires of a weak electrostatic septum, and then being diverted across the thick current-carrying conductor of a strong magnetic septum.

If the speed v_2 is large enough, then the change in displacement x from turn N to turn $N + 2$ is sufficient to jump across the wires of an electrostatic septum, as illustrated in Figure 10.5. Some protons will inevitably strike the wires, but most of them jump clear to receive an additional angular kick from the electric field outside the wires. The electrostatic force delivering this kick is relatively weak, since

$$E \ll cB \tag{10.20}$$

for typical values of E and B in the Lorentz equation

$$\vec{F} = q(\vec{E} + \vec{v} \times \vec{B}) \tag{10.21}$$

Nonetheless, an electrostatic septum has the overwhelming advantage of placing a much narrower strip of material in the way of the beam. Electrostatic wires can be as small as 50 μm, while the copper conductors of a magnetic septum are typically a few millimetres wide. Much weaker octupoles efficiently drive almost all of the protons across an electrostatic septum, compared to a scheme with a primary magnetic septum.

Exercises

10.1 Analyse the tracking data shown in the figure, from a RHIC simulation. The 9 dots record the shifted tunes (Q_x, Q_y) of a set of on-momentum test particles, launched with horizontal amplitudes ranging from 1 mm to 9 mm, at a location with $\beta_x = 40$ m.

a) Plot Q_x and Q_y versus J_x, after estimating tunes from the figure.
b) What is the simplest fit to the tune versus action data?
c) What is the simplest and most likely dominant nonlinearity?

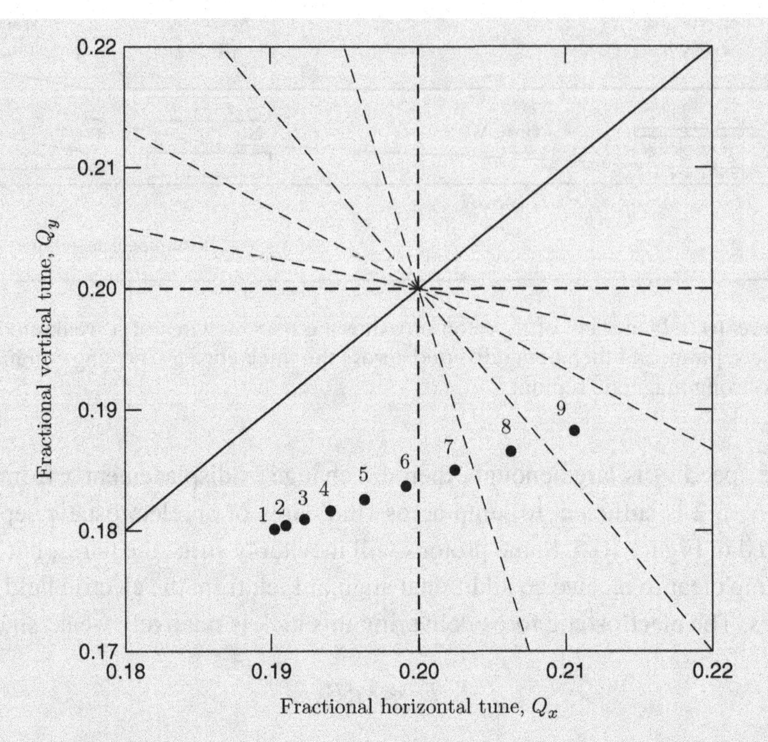

10.2 If a single dodecapole (12-pole) magnet delivers an angular kick

$$\Delta x' = -g_{12} x^5 \tag{10.22}$$

and causes normalised phase space detuning

$$Q = Q_0 + A\, g_{12} a^B \tag{10.23}$$

what are the values of the coefficients A and B?

10.3 How are the amplitudes u_k in Equation 10.10 related to the horizontal subset of amplitudes V_{i0k0} in Equation 10.14?

10.4 The *horizontal smear* of a trajectory is a dimensionless quantity S that measures the relative depth of the turn-by-turn fluctuations of the amplitude. Thus $S = 0$ for purely linear motion, and $S \leq 0.1$ (say) for acceptable near-linear motion, where

$$S \equiv \sigma_{xx}^{1/2} \tag{10.24}$$

and the normalised covariance is

$$\sigma_{xx} = \frac{< a_x^2 >}{< a_x >^2} - 1 \tag{10.25}$$

Recalling that the normalised horizontal amplitude on turn t is related to the action through

$$a_{x,t}^2 = 2J_{x,t} \tag{10.26}$$

develop an expression to relate the horizontal smear to the Fourier amplitudes u_k defined in Equation 10.10.

10.5 Consider the two-turn Hamiltonian H_2 plotted in Figure 10.3.

a) What are the (approximate) ratios of $V_0 : V_2 : V_4$?
b) What are the phases ϕ_2 and ϕ_4?
c) How should octupoles be distributed around the accelerator to achieve those values: how many same-strength octupoles at locations with identical Twiss functions, spaced by how much in horizontal phase?

10.6 Consider the electrostatic and magnetic septa sketched in Figure 10.6.

a) What are typical realistic values of E and B?
b) What is a typical ratio of electromagnetic forces, $E/(cB)$, for fully relativistic particles?
c) How small must the kinetic energy of a proton (or an electron) be, in order for electrostatic optics to be competitive with magnetic optics?

11

Synchrotron Radiation – Classical Damping

So far it has been implicitly assumed that charged particle motion in an accelerator is conservative, so that the amplitude and action of each particle remain constant (except for adiabatic damping during acceleration). However, a charged particle that accelerates transversely radiates photons, through the phenomenon of synchrotron radiation. The classical effect of this radiation is (usually) to damp the motion of each particle, so that amplitude and action decrease, and the beam size shrinks. But also, the quantum nature of random photon emission excites each particle, increasing amplitude, action and beam size.

Electrons in a storage ring quickly come to a dynamical equilibrium – the horizontal and longitudinal emittances stabilise at natural values with characteristic times that are usually very short, typically milliseconds. Protons, only about 2,000 times more massive than electrons, do not radiate at all for most practical purposes; their characteristic relaxation times are very long, typically many days, except for extremely high-energy accelerators like the LHC.

The classical aspects of synchrotron radiation are explored first in this chapter, before turning to quantum effects in Chapter 12.

11.1 Spectrum and Distribution Pattern

Radiation by an accelerating charged particle was understood in a classical context in the nineteenth century, long before the quantum nature of light and special relativity were understood. Larmor showed that a particle with charge q and acceleration a_c radiates with a power

$$P = \frac{1}{6\pi\epsilon_0} \cdot \frac{q^2 a_c^2}{c^3} \tag{11.1}$$

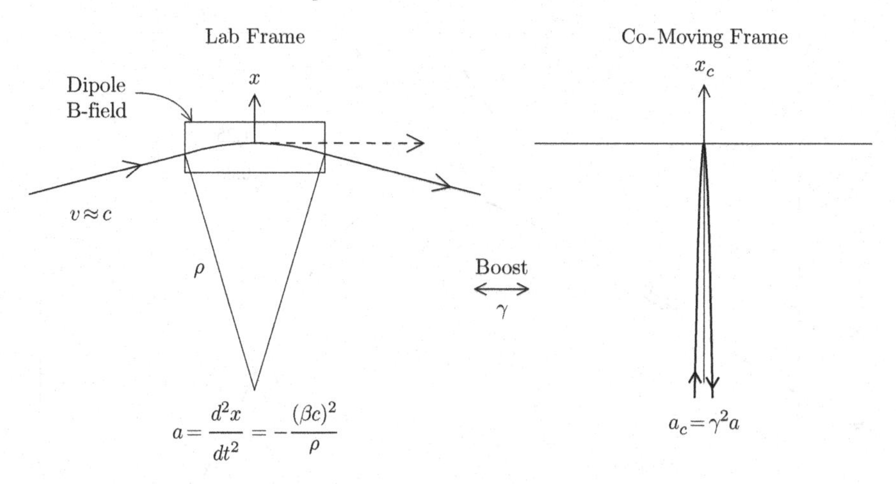

Figure 11.1 Larmor acceleration in a dipole magnet, as seen in the laboratory and tangentially co-moving frames that are related by a Lorentz boost of strength γ. The electron beam acceleration is perpendicular to the direction of motion.

where the permittivity of free space is

$$\epsilon_0 = \frac{1}{\mu_0 c^2} = 8.854 \times 10^{-12} \text{ [F/m]} \tag{11.2}$$

the permeability of free space has the defined value

$$\mu_0 \equiv 4\pi \times 10^{-7} \text{ [H/m]} \tag{11.3}$$

and c is the speed of light. It turns out that radiated power is a Lorentz invariant, so that boosting to the laboratory frame as in Figure 11.1 gives

$$P = \frac{1}{6\pi\epsilon_0} \cdot \frac{q^2 c}{\rho^2} \cdot \gamma^4 \tag{11.4}$$

where ρ is the bending radius. This is more conveniently written

$$P = \frac{1}{6\pi\epsilon_0} \cdot \frac{e^4}{m^4 c^5} \cdot B^2 E^2 \tag{11.5}$$

where m and E are the mass and total energy of the particle, and B is the field of the bending magnet through which it passes. The fourth power of mass makes the denominator 13 orders of magnitude larger for protons than for electrons!

The synchrotron radiation spectrum $\mathcal{P}(\omega)$ gives the total power P after integrating over the angular frequency ω. It has the universal shape

$$\mathcal{P}(\omega) = \frac{P}{\omega_c} S(\omega/\omega_c) \tag{11.6}$$

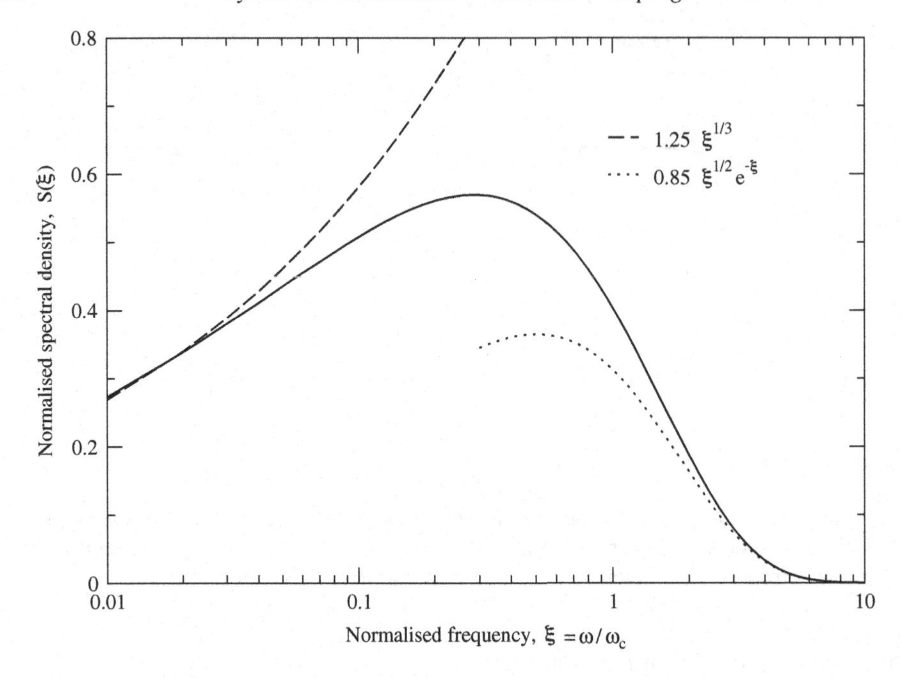

Figure 11.2 The universal shape of the synchrotron radiation spectrum $S(\omega/\omega_c)$, with a peak near the characteristic frequency ω_c.

shown in Figure 11.2, where

$$S(\xi) = \frac{9\sqrt{3}}{8\pi}\,\xi \int_{\xi}^{\infty} K_{5/3}(\bar{\xi})\, d\bar{\xi} \tag{11.7}$$

and $K_{5/3}$ is a modified Bessel function [21, 46]. The characteristic angular frequency

$$\omega_c = \frac{3}{2}\frac{c\gamma^3}{\rho} \tag{11.8}$$

is near the peak of the power spectrum, moving to higher frequencies in inverse proportion to the bending radius ρ, and cubically with γ. The corresponding characteristic wavelength is conveniently given for electrons by

$$\lambda_c\,[\mathrm{m}] = \frac{1.86 \times 10^{-9}}{B\,[\mathrm{T}]\,E^2\,[\mathrm{GeV}^2]} \tag{11.9}$$

Half of the total radiated power is emitted above the characteristic value, and half below.

The shape of the power distribution in the plane of a storage ring is

$$\frac{dP}{d\Omega} = \frac{P_0}{(1-\beta\cos\theta)^3}\cdot\left[1 - \frac{\sin^2\theta}{\gamma^2(1-\beta\cos\theta)^2}\right] \tag{11.10}$$

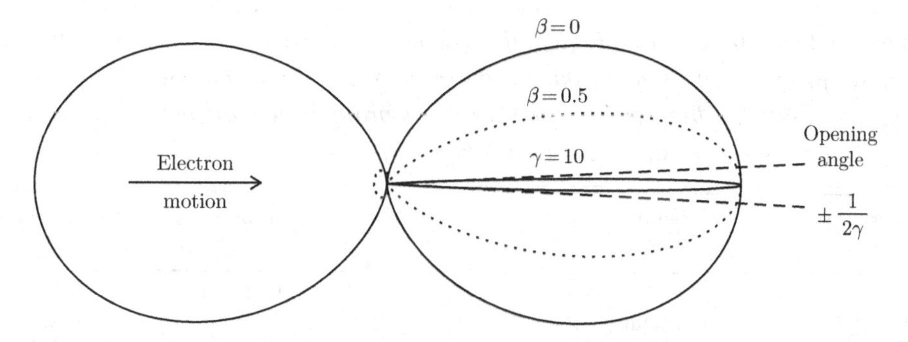

Figure 11.3 The synchrotron radiation pattern for three values of β and γ: non-relativistic, intermediate and relativistic. The total opening angle is $1/\gamma$ when $\gamma \gg 1$.

where θ is the angle from the direction of motion. Non-relativistic electrons with $\beta \ll 1$ broadcast with the classical two-lobed dipole radiation pattern, as shown in Figure 11.3. In the relativistic limit $\gamma \gg 1$ the shape becomes

$$\frac{dP}{d\Omega} \approx \frac{P_1}{(1+\gamma^2\theta^2)^3} \cdot \left[1 - \frac{4\gamma^2\theta^2}{(1+\gamma^2\theta^2)^2}\right] \tag{11.11}$$

so that the synchrotron radiation shines ahead of the electrons in a searchlight beam with a narrow opening angle of $\pm 1/2\gamma$, as illustrated in Figure 11.3. Electrons in a 1 GeV storage ring have $\gamma \approx 2,000$, so that the synchrotron radiation is naturally collimated within a 1 mrad angle. Such bright light sources began to be used in the 1970s, in the parasitic application of what were then primarily high-energy physics accelerators. The ensuing historical trend towards a global proliferation of purpose-built synchrotron light facilities is summarised in Table 11.1. These light sources continue to develop, becoming ever brighter, even while the storage energy has stabilised at an optimum value of about 3 GeV.

11.2 Energy Loss Per Turn and Longitudinal Damping

The total energy loss per particle per turn is in general

$$U = \oint P(s)\, \frac{ds}{c} \tag{11.12}$$

with an ideal design value

$$U_0 = C_g E_0^4 \, \frac{C}{2\pi} \, \langle G^2 \rangle \tag{11.13}$$

for a particle with energy E_0 circulating in an error-free storage ring of circumference C, where angle brackets $\langle\rangle$ denote a ring average, and $G = 1/\rho$ measures the local geometric bending strength. The convenient constant

Table 11.1 *A partial chronological list of synchrotron radiation rings, showing the general progression from the parasitic use of synchrotrons and electron–positron colliders built for high-energy physics, to a proliferation of purpose-built high performance electron storage rings (Wikipedia).*

Operating years	Name	Location	Energy [GeV]
1961-	SURF	Gaithersburg	0.18
1968-87	Tantalus	Madison	0.24
1972-75	Solidi Roma	Frascati	1.0
1973-88	ACO	Orsay	0.54
1973-	SSRL	Stanford	3.0
1974-93	DORIS	Hamburg	5.0
1974-	INS-SOR	Tokyo	0.3
1979-	CHESS	Ithaca	5.5
1981-2006	DCI	Orsay	1.0
1997-	HSRC	Hiroshima	0.7
1982-2014	NSLS-I	Upton	2.8
1982-	Photon Factory	Tsukuba	2.5
1986-	MAX-I	Lund	0.55
1987-2006	Super-ACO	Orsay	0.8
1991-	BSRF	Beijing	2.5
1991-	NSRL	Hefei	0.8
1992-	ESRF	Grenoble	6.0
1993-	ALS	Berkeley	1.9
1993-2012	DORIS III	Hamburg	5.0
1993-	ELETTRA	Trieste	2.4
1995-	APS	Lemont	1.9
1997-	HSRC	Hiroshima	0.7
1997-	LNLS	Campinas	1.4
1997-	Spring-8	Sayo	8.0
1998-	BESSY II	Berlin	1.7
1999-	Indus 2	Indore	2.5
1999-	SIBIR	Moscow	2.5
2000-	ANKA	Karlsruhe	2.5
2001-	SLS	Villigen	2.8
2004-	CLS	Saskatoon	2.9
2004-	SLRI	Suranari	1.2
2006-	Australian Synchrotron	Clayton	3.0
2006-	Diamond	Abingdon	3.0
2006-	SOLEIL	Orsay	3.0
2007-	SSRF	Shanghai	3.5
2009-	PETRA III	Hamburg	6.5
2010-	ALBA	Barcelona	3.0
2015-	NSLS-II	Upton	3.0
2015-	Taiwan PS	Hsinchu	3.0
2016-	MAX-IV	Lund	3.0

$$C_g = \frac{4\pi}{3} \frac{r_0}{(mc^2)^3} \tag{11.14}$$

is proportional to the classical radius of the particle

$$r_0 = \frac{q^2}{4\pi\,\epsilon_0\, mc^2} \tag{11.15}$$

so that $C_g \sim q^2/m^4$ and

$$\begin{aligned} C_g &= 8.85 \times 10^{-5}\ [\mathrm{m\,GeV^{-3}}] \quad \text{electrons} \\ &= 7.78 \times 10^{-18}\ [\mathrm{m\,GeV^{-3}}] \quad \text{protons} \end{aligned} \tag{11.16}$$

This underlines why protons do not radiate, for most practical purposes.

All dipoles have the same bending radius ρ in the simplified case of an isomagnetic ring, although their lengths may vary. In this case

$$\langle G^n \rangle = \frac{\oint G^n\, ds}{\oint ds} = \frac{1}{R\rho^{n-1}} \tag{11.17}$$

for any exponent n, where the average radius is

$$R = \frac{C}{2\pi} \tag{11.18}$$

and the energy loss per particle per turn in an isomagnetic ring is simply

$$U_0 = \frac{C_g E_0^4}{\rho} \tag{11.19}$$

For electrons

$$U_0\,[\mathrm{keV}] = 88.5\, \frac{E_0^4\,[\mathrm{GeV}^4]}{\rho\,[\mathrm{m}]} \tag{11.20}$$

so that U_0 = 0.29 MeV in the full-energy electron injector to NSLS-II, with E_0 = 3.0 GeV, B = 0.4 T and ρ = 25 m. By contrast, U_0 = 3.5 MeV in the extreme case of a speculative 50 TeV proton collider with ρ = 14 km.

Longitudinal motion was previously discussed for conservative particles that do not radiate in Section 4.3. The slip factor denoted by η_s in the proton accelerator community is usually called the (momentum) compaction factor α in the electron community, so that the evolution of the longitudinal displacement z is rewritten

$$z_{n+1} = z_n \; - \; \alpha C \cdot \delta_n \tag{11.21}$$

where, in the relativistic regime $\gamma \gg 1$,

$$\alpha \; = \; \langle \eta G \rangle \tag{11.22}$$

in direct analogy with Equations 4.25 and 4.27.

The turn-by-turn evolution of the off-momentum parameter δ must be modified from the conservative case, to include the continual replacement of electron energy lost to synchrotron radiation. It is convenient to introduce the synchronous phase ϕ_s through the definition

$$U_0 = eV \sin \phi_s \tag{11.23}$$

where V is the RF voltage amplitude summed over all RF stations, and then to write the motion as

$$\delta_{n+1} = \delta_n + \frac{1}{E_0}[eV(\sin \phi - \sin \phi_s) - (U(\delta_n) - U_0)] \tag{11.24}$$

where ϕ is the RF phase and $E = pc$ in the relativistic limit, so that

$$\delta = \frac{\Delta p}{p_0} \approx \frac{\Delta E}{E_0} \tag{11.25}$$

A particle launched with no momentum offset at the synchronous phase, at $(\delta, \phi) = (0, \phi_s)$, remains there forever, as illustrated in Figure 11.4. Stable motion requires that $eV > U_0$.

If the longitudinal oscillations are slow enough – if the synchrotron tune Q_s is sufficiently small – then the difference equations 11.21 and 11.24 in turn number n can legitimately be translated into differential equations in time t. In the relativistic limit, phase slippage becomes

$$\frac{d\phi}{dt} = -\alpha \omega_{RF} \cdot \delta \tag{11.26}$$

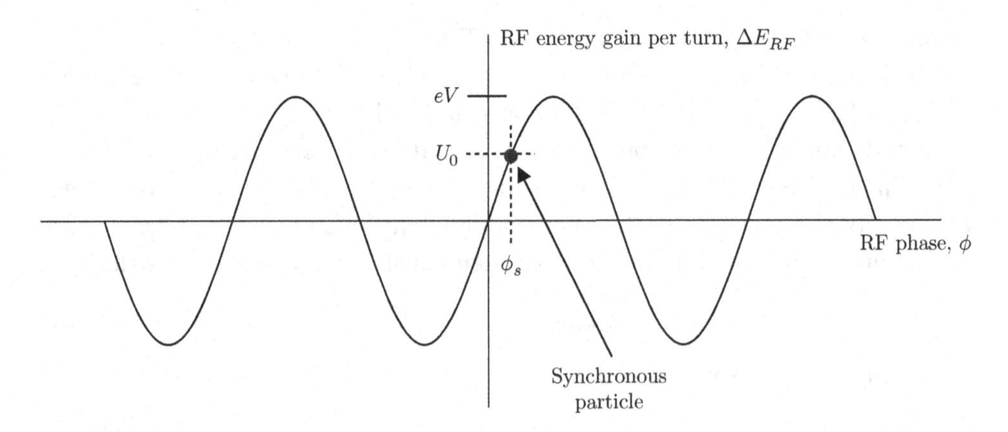

Figure 11.4 Energy gain from the RF system for an electron with phase ϕ that radiates $eV \sin \phi_s$ per turn, where ϕ_s is the synchronous phase.

while momentum evolution becomes

$$\frac{d\delta}{dt} = \frac{1}{TE_0}\left[eV(\sin\phi - \sin\phi_s) \;-\; (U(\delta_n) - U_0)\right] \tag{11.27}$$

where the angular RF frequency is

$$\omega_{RF} = 2\pi f_{RF} \tag{11.28}$$

and the revolution period

$$T = \frac{C}{c} \tag{11.29}$$

is the nominal time taken for one turn. Linearising Equations 11.26 and 11.27 and combining them into a single approximate second-order differential equation gives

$$\frac{d^2\delta}{dt^2} + \frac{2}{\tau_s}\frac{d\delta}{dt} + \left(\frac{2\pi Q_s}{T}\right)^2 \delta = 0 \tag{11.30}$$

with a damped solution

$$\delta = \delta_0\, e^{-t/\tau_s} \cos\left(2\pi Q_s \frac{t}{T}\right) \tag{11.31}$$

that is valid for small values of δ_0, the initial oscillation amplitude. The longitudinal damping time

$$\tau_s = \frac{2T}{dU/dE|_0} \tag{11.32}$$

depends on the slope of U at the nominal energy E_0.

The variation of the energy loss per turn with momentum is vital. $U(\delta)$ varies quadratically to lowest order in δ, as sketched in Figure 11.5, even in the ideal case when no errors are present, since

$$U(\delta) \;\sim\; \oint B^2 E^2\, ds \tag{11.33}$$

while

$$B(\delta) \;\sim\; G + K\,(\eta\delta + x_{co}) \tag{11.34}$$

where K is the local quadrupole strength, η is the dispersion, x_{co} is the closed orbit error and

$$E(\delta) \approx E_0(1+\delta) \tag{11.35}$$

In some perturbed conditions the slope $dU/d\delta$ can become negative, and longitudinal oscillations are antidamped. Larger rings are more susceptible to such errors.

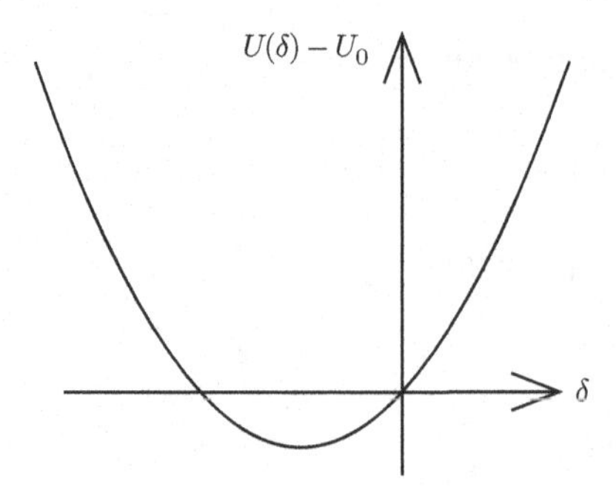

Figure 11.5 Variation of energy loss per turn U with momentum offset δ. The slope $dU/d\delta$ is naturally positive at $\delta = 0$ under ideal design conditions, and so longitudinal oscillations are damped.

A quantitative discussion of the longitudinal damping time awaits a discussion of transverse betatron oscillation damping, since (it turns out) the damping rates in all three dimensions have a constant sum.

11.3 Continuous Acceleration

But first it is natural to return, briefly, to the conservative situation in which $U(\delta)$ is zero, but in which ϕ_s remains non-zero, so that an ideal synchronous particle picks up more and more energy – the beam continually accelerates. In this case the acceleration Hamiltonian

$$H(\phi, \delta) = \frac{1}{2}\alpha\omega_{RF}\,\delta^2 - \frac{eV}{TE_0}\,(\cos\phi + \phi\sin\phi_s) \tag{11.36}$$

reproduces Equations 11.26 and 11.27 through the canonical equations

$$\begin{aligned} \frac{d\delta}{dt} &= \frac{\partial H}{\partial \phi} \\ \frac{d\phi}{dt} &= -\frac{\partial H}{\partial \delta} \end{aligned} \tag{11.37}$$

Motion along the contours of the Hamiltonian is illustrated in Figure 11.6. For example, a particle with an initially large positive momentum offset δ is untrapped. At first it traverses a contour towards lower ϕ on the high-δ side of the valley, before passing between two lakes and then traversing towards larger ϕ on the low-δ side. By contrast, a synchronous particle with $(\delta, \phi) = (0, \phi_s)$ at the centre of one of the lakes of stability remains there forever. Only particles that are trapped in one of the lakes – in an RF bucket – are successfully accelerated. Untrapped particles eventually hit the vacuum chamber wall, when δ becomes sufficiently negative.

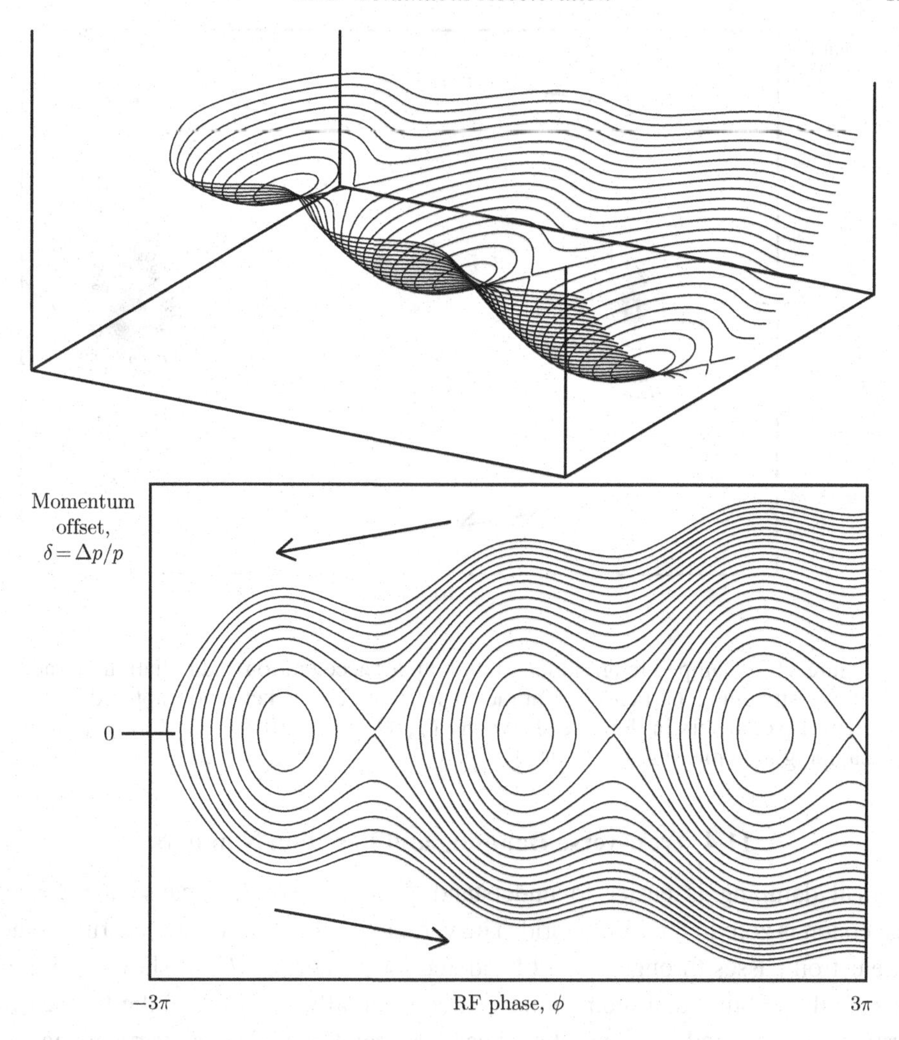

Figure 11.6 Hamiltonian contours followed during conservative longitudinal motion under constant acceleration, with a synchronous phase of $\phi_s = 0.4$ radians. Some particles are trapped in the periodic lakes of stability on the valley floor, while others are untrapped and stream continuously along the sides of the valley.

Particles no longer follow Hamiltonian contours when U and $dU/d\delta|_0$ are turned back on, as illustrated in Figure 11.7, which shows the motion of a set of particles launched under one particular trio of control parameters (ϕ_s, Q_s, τ_s). In addition to replacing radiated energy, the RF system may also be accelerating the beam. As before, some particles squeeze between the RF buckets to be lost at negative δ when they strike a physical aperture limit. Unlike before, some (apparently) untrapped particles are attracted into an RF bucket, and are captured. There is a mathematical connection, here, to the world of non-conservative strange attractors.

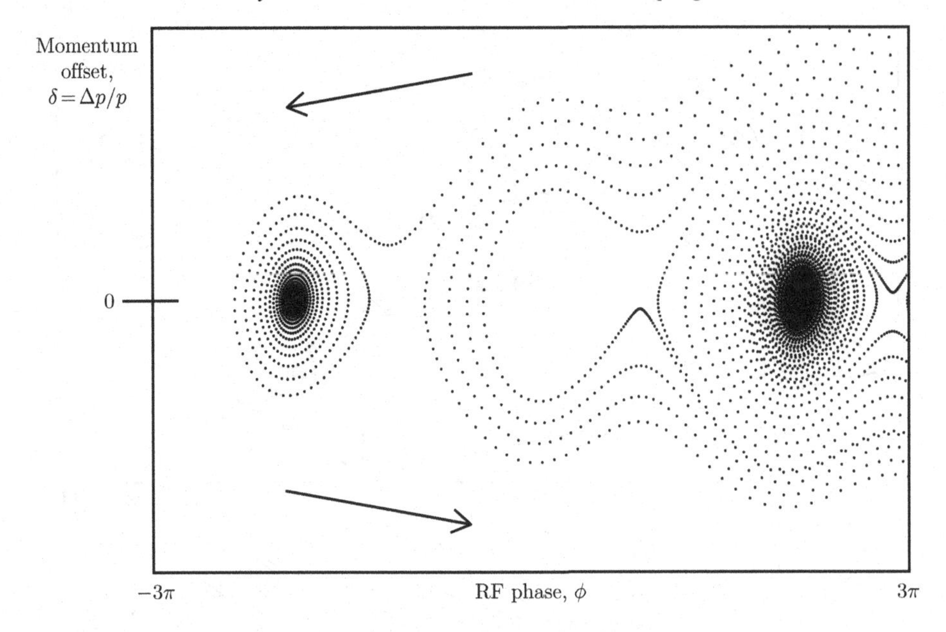

Figure 11.7 Damped longitudinal motion under acceleration and radiation. Some of the streaming particles are attracted into an RF bucket, and captured. The control parameters in the case shown are $(\phi_s, Q_s, \tau_s) = (0.4, 0.02, 300)$, with the damping time measured in turns.

11.4 Transverse Damping and Partition Numbers

Vertical damping is easier to understand than horizontal, because the vertical dispersion is zero in the ideal world. The vertical momentum vector $\vec{p}$ shrinks when an electron passes through a short bend and loses energy ΔU, as shown in Figure 11.8. If the longitudinal momentum that is eventually replaced by the lumped RF system is considered to be distributed bend-by-bend, then the new vertical angle on leaving the dipole is connected to the old angle on entrance through

$$y'_{NEW} = \frac{(1 - \Delta U/E)p_{\perp}}{p_0} = \left(1 - \frac{\Delta U}{E}\right) \cdot y'_{OLD} \tag{11.38}$$

Vertical betatron oscillations therefore damp according to

$$y = a_y e^{-t/\tau_y} \cos\left(2\pi Q_y \frac{t}{T}\right) \tag{11.39}$$

where the vertical damping time τ_y is equal to the characteristic damping time τ_0, defined by

$$\tau_0 \equiv 2T \frac{E_0}{U_0} \tag{11.40}$$

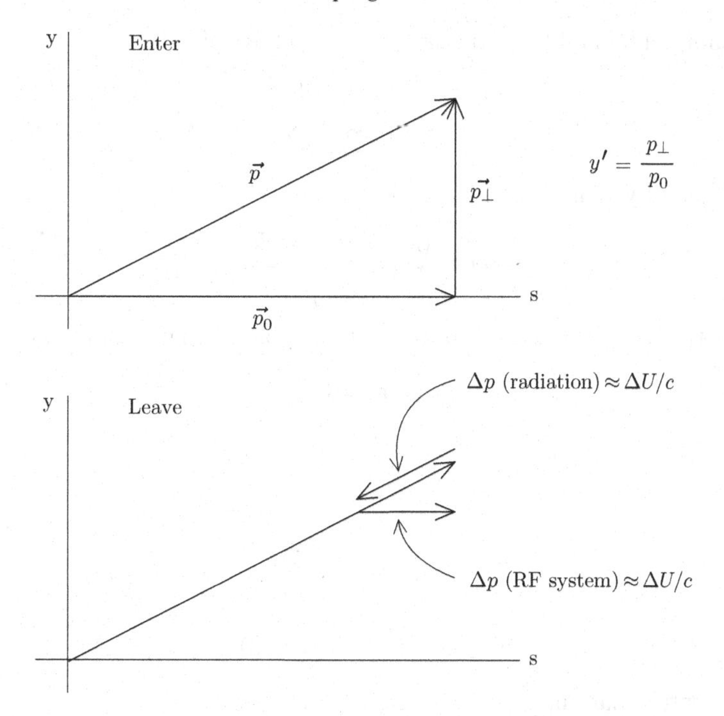

Figure 11.8 Vertical damping due to motion through a single dipole. The momentum vector $\vec{p}$ shrinks by $\Delta U/c$ during the passage, with no change in direction. In contrast, the momentum derived (eventually) from the RF system is replaced parallel to the s-axis. Hence the vertical angle y' decreases.

Curiously, the characteristic time is twice as long as it would take for a particle to radiate all of its energy E_0, if it continued to radiate U_0 on every turn! Also, note that τ_0 does not depend on the slope $dU/d\delta$ – the vertical damping time has no significant sensitivity to radiation in quadrupole magnets, for example due to closed orbit errors.

The horizontal damping time can be calculated in a similar fashion, but with the complication that the total horizontal displacement and angle have off-momentum contributions through the horizontal dispersion, and its slope. For present purposes it is sufficient to state that the three damping times are related to each other through

$$\tau_{x,y,s} = \frac{\tau_0}{J_{x,y,s}} \tag{11.41}$$

where the partition numbers J are simply related through the sum

$$J_x + J_y + J_s = 4 \tag{11.42}$$

The vertical partition number is

$$J_y = 1 \tag{11.43}$$

while the horizontal and longitudinal are related through

$$\begin{aligned} J_x &= 1 - \mathcal{D} \\ J_s &= 2 + \mathcal{D} \end{aligned} \tag{11.44}$$

where the quantity $\mathcal{D}$ is given in general by

$$\mathcal{D} = \frac{\langle \eta G^3 \rangle + 2 \langle \eta G K \rangle}{\langle G^2 \rangle} \tag{11.45}$$

In an error-free separated function ring with no combined function magnets

$$GK = 0 \tag{11.46}$$

so that ideally

$$\mathcal{D} = \frac{\langle \eta G^3 \rangle}{\langle G^2 \rangle} \sim \frac{\eta}{\rho} \ll 1 \tag{11.47}$$

and

$$(J_x, J_y, J_s) \approx (1, 1, 2) \tag{11.48}$$

to a good approximation.

$\mathcal{D}$ easily becomes of order one in a combined function ring, even with ideal optics, leading to longitudinal or horizontal antidamping. This is why electron accelerators with storage times longer than the characteristic damping time τ_0 always use separated function optics. Hadron accelerators are free to use combined function optics – some do, some do not. The local quadrupole strength K feeds down into the bending strength G when closed orbit errors are present, as described by Equation 11.34, and $\mathcal{D}$ can become comparable to one in large electron accelerators, threatening stability even with separated function optics [13, 38].

If motion in all three dimensions is damped, what prevents electron beams from becoming vanishingly small? So far the Planck constant h has not entered the conversation at all. However, its non-zero value is crucial in parameterising the strength of the quantum excitation effects that are in dynamic competition with the classical damping described in this chapter.

Exercises

11.1 What is the ratio of the power coefficients P_0 and P_1 in Equations 11.10 and 11.11?

11.2 The Lorentz transformations of electromagnetic fields are

$$\vec{E}'_{\perp} = \gamma \left(\vec{E}_{\perp} + \vec{v} \times \vec{B}_{\perp} \right) \tag{11.49}$$

$$\vec{E}'_{\parallel} = \vec{E}_{\parallel} \tag{11.50}$$

$$\vec{B}'_\perp = \gamma\,(\vec{B}_\perp - \vec{v} \times \vec{E}_\perp) \qquad (11.51)$$

$$\vec{B}'_\parallel = \vec{B}_\parallel \qquad (11.52)$$

where $\parallel$ labels the component of the field parallel to the boost velocity $\vec{v}$ and $\perp$ labels the perpendicular. Compute the electromagnetic fields of a singly charged particle travelling at relativistic velocity $\vec{v}$.

11.3 Consider 50 GeV electrons in LEP *(requiesce in pace)*, circulating in a nearly circular ring with a total circumference of about 27 km. Consider also a 5 TeV electron storage ring that is built around the earth's equator. For each accelerator:

a) How much energy is radiated per electron per turn?
b) What is the characteristic damping time τ_0, in turns and in seconds?
c) What is the bending field?

11.4 In RHIC, 55 bunches of 10^9 gold ions with $(Z, A) = (79, 197)$ circulate in each ring, at a top energy of $\gamma = 108$. The circumference is 3833 m and the main arc dipole bending radius is $\rho = 243$ m.

a) How much energy is radiated per gold ion per turn?
b) What is the characteristic damping time τ_0, in turns and in seconds?
c) The synchrotron radiation power is a serious cryogenic load if it exceeds about 1 W/m. Is it a problem?
d) If 360 bunches of 2×10^{11} 10 GeV electrons circulate in a new ring in the existing RHIC tunnel, how much energy is radiated per electron per turn, and what is the characteristic damping time τ_0?

11.5 A Future Circular Collider (FCC) might circulate 50 TeV protons in a 100 km circumference tunnel with a main arc dipole bend radius of 11 km.

a) What is the arc dipole bending field?
b) What is the critical energy of photons radiated in the dipoles?
c) What is the total energy lost per turn, per proton?
d) If each proton beam has a current of 0.5 A, what is the total synchrotron radiation power, per ring?
e) Assuming that cryogenic refrigerators operate with a Carnot efficiency of 20%, how much 'wall-plug' power would be required if the synchrotron radiation were absorbed at a temperature of 4 K?

11.6 Consider muon storage rings.

a) What is the total power radiated in an isomagnetic muon storage ring of radius ρ with a beam current of I?

b) How much power is emitted in a 30 GeV ring with $\rho = 250$ m that stores 1 A of electrons, or 1 A of muons?

c) What is the natural beam current lifetime decay for 30 GeV muons?

d) What is the heat load due to the decay of 1 A of 30 GeV muons?

11.7 (See also Exercise 12.5.) Consider the interplay between damping and resonances. Modify a Hénon map simulation to include a turn-by-turn amplitude decrement of $\Delta = 1/\tau$.

a) With $\Delta = 0$ select Q so that a chain of resonance islands appears, centred at a moderate phase space amplitude. Increase Δ and observe that test particles can get trapped in an island, spiralling towards its centre, if Δ is not too big. This phenomenon of electron resonance trapping was observed in synchrotron light monitors as early as 1968, in Novosibirsk [26].

b) How does the critical value for trapping Δ_c depend on Q for a particular island chain? What is your interpretation?

11.8 (See also Exercises 4.4 and 5.8.) RHIC accelerates fully stripped gold ions in an RF system with a harmonic number of $h = 360$ and with a typical voltage of $V_{RF} = 300$ kV. Assume that transition $\gamma_T = 22.89$, and ignore synchrotron radiation.

a) If $\phi_s = 0°$ before acceleration begins, when $\gamma = 10.4$, what is the synchrotron frequency (in Hz)?

b) For a synchronous phase of $\phi_s = 5.5°$, how much energy does the synchronous particle gain per turn?

c) How long does it take to accelerate to $\gamma = 107.4$? Assume that the phase jump at transition has been performed correctly – ignore it.

d) Plot the synchrotron frequency as a function of energy, up the acceleration ramp.

12

Synchrotron Radiation – Quantum Excitation

In the discussion of synchrotron radiation so far, it has only been necessary to call upon classical electrodynamics from the late nineteenth century, and special relativity from the early twentieth century. Now, however, it is necessary to admit that synchrotron light is emitted as photons with a characteristic energy of

$$u_c = \hbar\omega_c = \frac{3}{2}\frac{\hbar c\gamma^3}{\rho} \tag{12.1}$$

Random quanta excite the beam.

The reduced Planck constant, introduced here for the first time, has a small but non-zero value

$$\begin{aligned}\hbar &= 1.055 \times 10^{-34}\ [\text{J s}] \\ &= 6.582 \times 10^{-16}\ [\text{eV s}]\end{aligned} \tag{12.2}$$

so that, for example, 3 GeV electrons bent by a dipole field of 0.4 T with a radius of 25 m typically emit 2.4 keV photons. More conveniently

$$u_c\,[\text{keV}] = 0.665\, E^2\,[\text{GeV}^2]\, B\,[\text{T}] \tag{12.3}$$

The total energy loss per electron per turn is 0.29 MeV in an isomagnetic ring made from such dipoles, according to Equation 11.19, corresponding to about 0.8 photons per meter per turn. Synchrotron radiation is lumpy!

The total number of photons emitted per second is

$$N = \int_0^\infty n(u)\, du \tag{12.4}$$

where the quantum spectrum $n(u)$ is related to the classical power spectrum $\mathcal{P}(\omega)$ introduced in Equation 11.6 through

$$\mathcal{P}(\omega) = u\, n(u)\frac{du}{d\omega} = \hbar u\, n(u) \tag{12.5}$$

Quantum excitation has a strength that is usefully measured (it will be shown) by

$$N\langle u^2\rangle = \int_0^\infty n(u)\, u^2\, du \tag{12.6}$$
$$= \frac{55}{24\sqrt{3}} r_0 \hbar m c^4 \frac{\gamma^7}{\rho^3}$$

Thus, the strength of quantum excitation scales with the seventh power of the energy γ and the third power of G, the local bending strength.

12.1 Energy Spread

When a single photon of energy u is emitted, the *exact* change in longitudinal amplitude, from A_0 to A_1, is given by

$$A_1^2 = A_0^2 + u^2 - 2A_0 u \cos(\phi) \tag{12.7}$$

as shown in Figure 12.1. Averaging over phase ϕ leads to

$$\langle \Delta A^2\rangle = \langle A_1^2 - A_0^2\rangle = \langle u^2\rangle \tag{12.8}$$

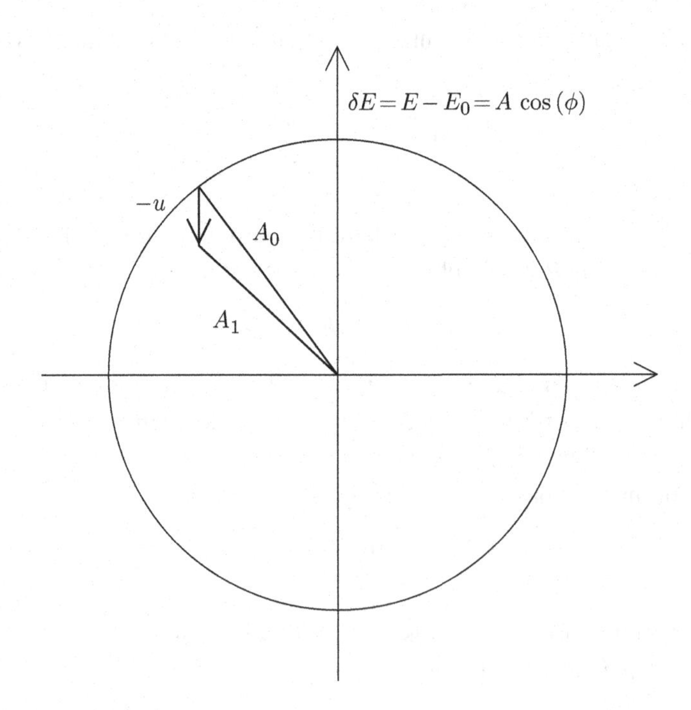

Figure 12.1 The change in longitudinal amplitude, from A_0 to A_1, when a photon of energy u is emitted.

and averaging over many photons confirms that the rate of random walk

$$\frac{d\langle A^2\rangle}{dt} = \left\langle \frac{dA^2}{dt} \right\rangle = N\langle u^2\rangle \tag{12.9}$$

is just the excitation strength introduced in Equation 12.6. $N\langle u^2\rangle$ represents *any* random source of energy excitation – instead of quantum emission it could equally well represent noise in the RF system of a hadron ring.

Turn-by-turn damping, as already described by Equation 11.31, is readily rewritten in the same format

$$\frac{d\langle A^2\rangle}{dt} = -2\frac{\langle A^2\rangle}{\tau_s} \tag{12.10}$$

so that when $d\langle A^2\rangle/dt = 0$ at equilibrium between damping and excitation

$$\langle N\langle u^2\rangle\rangle_s - 2\frac{\langle A^2\rangle}{\tau_s} = 0 \tag{12.11}$$

where the excitation term has been averaged around the entire ring. The natural energy spread σ_E of the beam is therefore given by

$$\sigma_E^2 = \frac{1}{2}\langle A^2\rangle = \frac{\tau_s}{4}\langle N\langle u^2\rangle\rangle_s \tag{12.12}$$

and the natural relative energy spread is more conveniently written

$$\left(\frac{\sigma_E}{E_0}\right)^2 = \frac{C_q}{J_s}\gamma^2\frac{\langle G^3\rangle}{\langle G^2\rangle} \tag{12.13}$$

where the constant C_q – a partner to C_g defined in Equation 11.14 – is

$$\begin{aligned} C_q &= \frac{55}{32\sqrt{3}}\frac{\hbar}{mc} \\ &= 3.84\times10^{-13}\ [\text{m}] \qquad \text{electrons} \\ &= 2.09\times10^{-16}\ [\text{m}] \qquad \text{protons} \end{aligned} \tag{12.14}$$

The relative energy spread depends (through $G(s)$) only on the geometric footprint of the accelerator. It is completely independent of the Twiss functions and dispersion!

In the simple case of an isomagnetic ring with a longitudinal partition number $J_s = 2$ the natural energy spread is just

$$\left(\frac{\sigma_E}{E_0}\right)^2 = \frac{C_q}{2}\frac{\gamma^2}{\rho} \tag{12.15}$$

Table 12.1 *The natural relative energy spread σ_E/E for a selection of accelerators, in the isomagnetic approximation. Electron energy spreads tend to be about 10^{-3}. Proton energy spreads (due to synchrotron radiation) are negligible, by comparison, even in the extreme case of the LHC.*

Accelerator	Species	Energy γ	Bend radius ρ [m]	Energy spread σ_E/E
NSLS-II Injector	e	6×10^3	25	5.3×10^{-4}
CESR	e	10^4	90	4.6×10^{-4}
LEP	e	10^5	3,400	7.5×10^{-4}
LHC	p	7×10^3	3,400	1.7×10^{-6}

Energy spreads for a typical and broad selection of reasonably isomagnetic electron storage rings are remarkably constant, of order 10^{-3}, as shown in Table 12.1. This reflects the empirical fact that γ^2/ρ is approximately constant for many electron rings.

The RMS bunch length is closely related to the energy spread. Combining Equation 5.47 with Equation 12.13 shows that in the relativistic limit $\gamma \gg 1$

$$\sigma_s = \frac{C}{2\pi}\frac{\alpha}{Q_s}\left(\frac{\sigma_E}{E_0}\right) \tag{12.16}$$

where α, the compaction factor, *does* depend on the optics, via dispersion.

12.2 Horizontal Emittance

The horizontal betatron displacement and angle also change when a photon of energy u is emitted, because the total horizontal displacement of a particle has both off-energy and betatron components

$$x_{TOT} = \eta\frac{\delta E}{E_0} + x_\beta \tag{12.17}$$

where η is the dispersion function. Figure 12.2 illustrates how betatron motion is excited according to

$$\begin{pmatrix} \Delta x_\beta \\ \Delta x'_\beta \end{pmatrix} = \begin{pmatrix} \eta \\ \eta' \end{pmatrix}\frac{u}{E_0} \tag{12.18}$$

In the absence of excitation events, the square of the normalised betatron amplitude

$$a_x^2 = \gamma_x x_\beta^2 + 2\alpha_x x_\beta x'_\beta + \beta_x x'^2_\beta \tag{12.19}$$

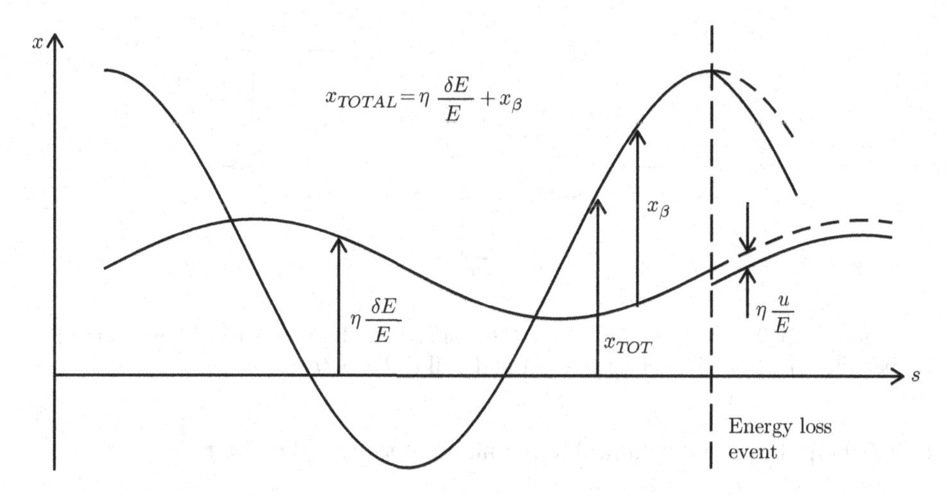

Figure 12.2 The excitation of the horizontal betatron displacement and angle in an energy loss event such as photon emission, or a noisy RF system.

remains constant as a function of azimuthal location s, and from turn to turn, as discussed in Section 5.1. But when a photon is emitted the amplitude changes according to

$$\Delta a_x^2 = \mathcal{H} \left(\frac{u}{E_0} \right)^2 \tag{12.20}$$

where the $\mathcal{H}$-function

$$\mathcal{H} = \gamma_x \eta^2 + 2\alpha_x \eta\eta' + \beta_x \eta'^2 \tag{12.21}$$

bears a strong similarity to the Courant–Snyder invariant introduced in Equation 5.4. However, $\mathcal{H}$ is *not* an invariant, although it is almost constant in some cases, for example across matched FODO cells. In general $\mathcal{H}(s)$ depends on the azimuthal location s through its direct dependence on the horizontal optical functions $(\gamma_x, \alpha_x, \beta_x, \eta, \eta')$. It depends on the footprint geometry G only indirectly, through the dispersion and its slope (η and η').

The rate of betatron excitation (averaged around the ring)

$$\frac{d\langle a_x^2 \rangle}{dt} = \frac{1}{E_0^2} \langle N \langle u^2 \rangle \mathcal{H} \rangle_s \tag{12.22}$$

competes with the horizontal damping

$$\frac{d\langle a_x^2 \rangle}{dt} = -2 \frac{\langle a_x^2 \rangle}{\tau_x} \tag{12.23}$$

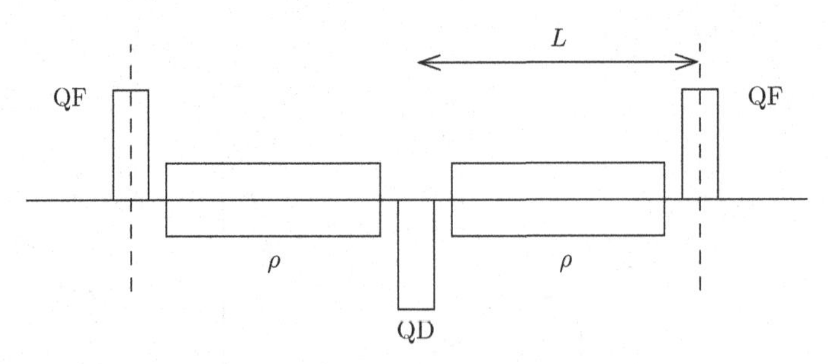

Figure 12.3 The basic FODO cell of an isomagnetic lattice with only one kind of bending magnet. The bend angle per half-cell is $\theta \approx L/\rho$.

so that at equilibrium the natural horizontal unnormalised emittance is

$$\epsilon_x \equiv \frac{1}{2}\left\langle a_x^2\right\rangle = \frac{\tau_x}{4E_0^2}\left\langle N\left\langle u^2\right\rangle \mathcal{H}\right\rangle_s \tag{12.24}$$

which is more conveniently written

$$\epsilon_x = \frac{C_q}{J_x}\gamma^2\frac{\left\langle G^3\mathcal{H}\right\rangle}{\left\langle G^2\right\rangle} \tag{12.25}$$

The natural horizontal emittance depends strongly on the lattice optics through $\mathcal{H}$, unlike the natural energy spread given by Equation 12.13, although the geometric and energy scaling is the same in both cases.

Again, the simple case of an isomagnetic ring is instructive, although it is also necessary to specify the repetitive cell structure of the lattice optics, so that $\mathcal{H}$ can be calculated. The simplest isomagnetic lattice is made entirely of matched FODO cells, with half-cell length L and bend angle θ, as shown in Figure 12.3. At the centres of the quadrupoles

$$\alpha_x = \eta' = 0 \tag{12.26}$$

so that $\gamma_x = 1/\beta_x$ and

$$\mathcal{H} = \frac{\eta^2}{\beta_x} \tag{12.27}$$

Thanks to the geometric scaling $\beta_x \sim L$ and $\eta \sim L\theta$ that has already been established in Sections 3.5 and 4.2, it is clear that

$$\mathcal{H} \sim L\theta^2 \sim \frac{L^3}{\rho^2} \tag{12.28}$$

although it is not immediately clear how much $\mathcal{H}$ varies within each FODO cell, or how $\mathcal{H}$ varies as a function of $\Delta\phi$, the phase advance per cell. $\mathcal{H}$ varies strongly

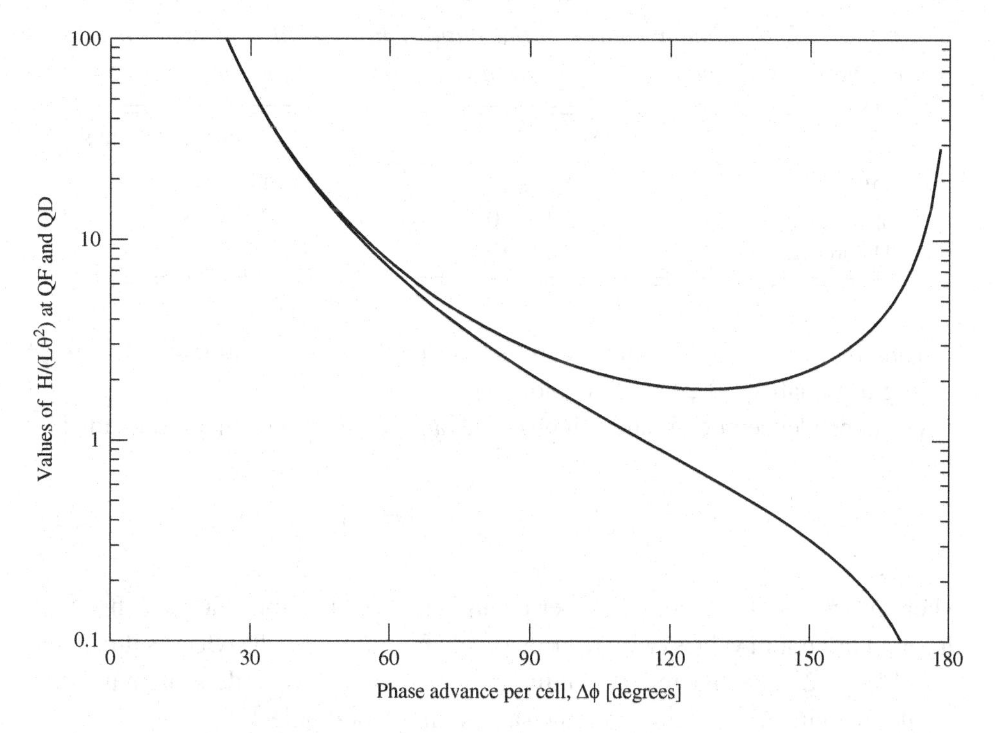

Figure 12.4 $\mathcal{H}$-function values at the QF and QD quadrupoles of the FODO cells of an isomagnetic lattice, as a function of the phase advance per cell. The two values are approximately equal for typical phase advance values of less than about 90 degrees.

with the phase advance per cell, as shown by the plot at the bottom of Figure 12.4, which draws on Equations 3.50 and 4.18. Nonetheless $\mathcal{H}$ is approximately equal at the QF and QD quadrupole centres (and implicitly also between them) for common values of $\Delta\phi$ in the range from 45 degrees to 90 degrees.

The natural horizontal unnormalised emittance scales like

$$\epsilon_x \sim \frac{C_q}{J_x} \gamma^2 \frac{L^3}{\rho^3} \tag{12.29}$$

where $J_x = 1$ is usually a good approximation, since

$$\frac{\langle G^3 \mathcal{H} \rangle}{\langle G^2 \rangle} \sim \frac{L^3}{\rho^3} \sim \theta^3 \tag{12.30}$$

Synchrotron light sources strive for brighter beams by reducing the natural emittances, for example by reducing the cell length $2L$. State-of-the-art synchrotron light sources often use non-FODO cells, designing them to have $\mathcal{H}$ small when G is large, and vice versa. No matter what cell style is chosen – for example,

Table 12.2 *Minimum values of the natural horizontal emittance form factor, F, for different cell styles in the isomagnetic approximation [35].*

Cell style	$F_{minimum}$	Accelerator example
FODO	7.3×10^{-4}	PEP
Chasman-Green	2.4×10^{-5}	NSLS VUV
(Theoretical)	7.8×10^{-6}	–

Chasman-Green, Double Bend Achromat or Triple Bend Achromat – the $\gamma^2\theta^3$ scaling of Equation 12.29 always holds true.

The dependence of emittance on phase advance per cell is explicitly included by writing

$$\epsilon_x\,[\mathrm{m}] = F(\Delta\phi)\,\frac{E^2\,[\mathrm{GeV}^2]}{J_x N_d^3} \tag{12.31}$$

where $N_d \sim 1/\theta$ is the total number of dipoles [35]. The form factor F has very different minimum values (with respect to $\Delta\phi$) for different cell styles, as illustrated in Table 12.2. The theoretical minimum emittance is two orders of magnitude smaller than in a FODO ring with the same number of dipoles!

Practical accelerators necessarily break the pure cell structure by including injection and extraction straights, insertion devices, RF systems and dispersion suppressors, etc. Other performance factors also intrude, such as dynamic aperture and real estate footprint. However, even a moderately detailed discussion of optimal synchrotron light source design is beyond the scope of this book!

12.3 Vertical Emittance

The vertical dispersion η_y is ideally zero if an accelerator is built in a horizontal plane, with no deliberate vertical bends. The natural vertical emittance is therefore also expected to be zero, since

$$\epsilon_y \;\sim\; \mathcal{H}_y \;\approx\; \frac{\eta_y^2}{\beta_y} \tag{12.32}$$

Reality intrudes, for example through vertical misalignments of quadrupoles and through roll misalignments of horizontal dipoles, both of which generate vertical dispersion. A second major source of vertical emittance is the linear coupling that inevitably exists between horizontal and betatron oscillations, due to skew quadrupole and solenoid magnetic field components. Linear coupling can generate vertical emittance, in addition to contributing in its own right to the vertical beam size.

Careful design, construction and operation can reduce the (effective) ratio of natural emittances to values

$$\frac{\epsilon_y}{\epsilon_x} \leq 10^{-2} \tag{12.33}$$

so that the real-space (x, y) profile of an electron beam is naturally flat if the horizontal and vertical β-functions are reasonably commensurate. Electron–positron collider rings usually go further, and arrange the optics so that

$$\beta_y^* \ll \beta_x^* \tag{12.34}$$

at the collision point, making the real-space profile even flatter there, and ameliorating the ravages of the beam–beam effect.

Exercises

12.1 Discuss the ‘empirical fact’ (mentioned near Equation 12.15) that γ^2/ρ is approximately constant over a broad range of electron accelerators.

a) Why might this be the case?
b) What are the consequences?
c) Under what circumstances might this scaling be broken?

12.2 Consider the dependence of the $\mathcal{H}$-function on the phase advance per FODO cell $\Delta\phi$, shown in Figure 12.4.

a) Which line traces the $\mathcal{H}$-value in QF, and which in QD?
b) Plot how $\mathcal{H}$ varies as a function of s between QF and QD quadrupoles, for a typical value of $\Delta\phi$. Is $\mathcal{H}$ approximately constant?
c) The horizontal and vertical phase advances per FODO cell can be quite different. How does the value of $\mathcal{H}$ at QF (or QD) vary in $(\Delta\phi_x, \Delta\phi_y)$ space?

12.3 A light source with a circumference of 176 m contains 8 identical Double Bend Achromat cells, with zero dispersion (and dispersion slope) at one end or the other of all 16 dipoles. Each dipole is 2.7 m long, and the beam energy is 2.5 GeV.

a) What is the characteristic energy of the photons radiated in the dipoles?
b) How much energy is radiated per turn, per electron?
c) What is the momentum compaction factor of the ring?
d) What are the damping times τ_x, τ_y and τ_s?
e) What is the natural horizontal emittance (approximately)?

12.4 The total horizontal beam size at a given location in an electron light source has two components:

$$\sigma_{\text{tot}}^2 = \sigma_{\text{beta}}^2 + \eta^2 \left(\frac{\sigma_E}{E_0}\right)^2 \tag{12.35}$$

where σ_{beta} is the betatron beam size, (σ_E/E_0) is the relative energy spread and η is the horizontal dispersion. Make reasonable approximations and assumptions as necessary.

a) Which of the two contributions is larger?
b) Is there a simple relationship between the two?

12.5 Exercise 11.7 added a radiation damping decrement Δ to the Hénon map. Now also add quantum excitation with a random walk step of RMS size α. It is necessary to track for many damping times $N \gg \tau = 1/\Delta$ for transients to die away. Use something like the following equations of motion:

$$\begin{aligned}
\begin{pmatrix} x \\ x' \end{pmatrix} &= \begin{pmatrix} c & s \\ -s & c \end{pmatrix} \begin{pmatrix} x \\ x' \end{pmatrix} \qquad c \equiv \cos(2\pi Q) \quad s \equiv \sin(2\pi Q) \\
x' &= x' - b_2\, x^2 \\
x &= x(1-\Delta) + \alpha G \\
x' &= x'(1-\Delta) + \alpha G
\end{aligned} \tag{12.36}$$

where b_2 is the strength of the sole sextupole, and G is a random number with a turn average of 0, and a standard deviation of 1.

a) With the sextupole turned on, observe how the beam size $\langle x^2 \rangle^{1/2}$ scales with α and Δ. Explain this mathematically.
b) Set $Q \approx 1/4$ so that four resonance islands appear. Launch particles in these islands. Under what conditions of (Δ, α) are particles trapped 'forever' in the islands?
c) How do the parameters b_2 and $(Q - 1/4)$ affect the trapping behaviour?

13

Linacs – Protons and Ions

Even the highest-energy proton linear accelerators, with kinetic energies of order 1 GeV, are not fully relativistic. Their $\beta = v/c$ values are significantly less than one. This is in stark contrast to electron linacs, in which $\beta \approx 1$ even for energies of only a few MeV. Because of this, the dominant dynamical issues are different in proton and electron linacs. Also, the menagerie of RF structures that accelerate and manipulate protons – each tuned for a different range of β values – is very diverse.

The technology map of Figure 13.1 shows some of the RF structures typically found in a modern large proton linac. Not shown are other common technologies, including Drift Tube Linacs (DTL), side coupled linacs, quarter wave resonators and double and triple spoke resonators. Fortunately an introduction to proton linac dynamics is possible without a detailed understanding of these diverse technologies, which are more fully discussed elsewhere [36, 57]. Instead it is possible to build on the generic discussion of pill-box and elliptical cavities in Chapter 7, adding only one instructive special case, the Radio Frequency Quadrupole (RFQ).

Even low energy proton rings – for example 250 MeV cancer therapy synchrotrons – need an injection linac, although these typically have maximum kinetic energies of less than 10 MeV. More interesting and comprehensive are high power linacs, such as the MW-class linacs listed in Table 13.1, which are used to generate neutrons for material science experimentation (ESS and SNS), to create rare isotopes for nuclear physics (FRIB), or to provide protons for high energy physics experimentation (PIP-II). In all four cases the hadron beam is absorbed in a target with a specialised design serving the need at hand.

This chapter does not discuss collective motion dynamics, or any of the related conceptual tools (such as wakefields and impedances), except to briefly introduce space charge as a phenomenon that is vital to linac design, and which connects to the beam-beam discussion in Chapter 15. Rather, the focus is on single-particle dynamics, especially the longitudinal phase space behaviour that dominates

Table 13.1 *Parameters of representative MW-class hadron linacs.*

Linac	Ion	Kinetic energy [GeV]	Beam power [MW]	Pulse current [mA]	Pulse length [ms]	Repetition rate [Hz]	Max RF freq. [MHz]
ESS	p	2.0	5.0	62	2.86	14	704
FRIB	p	0.61	0.4	0.66	–	–	322
	U	0.20[a]	0.4	0.0084	–	–	
PIP-II	H^-	0.8	1.2[b]	2	0.55[c]	20	650
SNS	H^-	0.94	1.4	27	0.97	60	805

[a] Kinetic energy for each of 238 nucleons.
[b] After accumulation and acceleration of half the beam to 120 GeV.
[c] Compatible with continuous waveform operation.

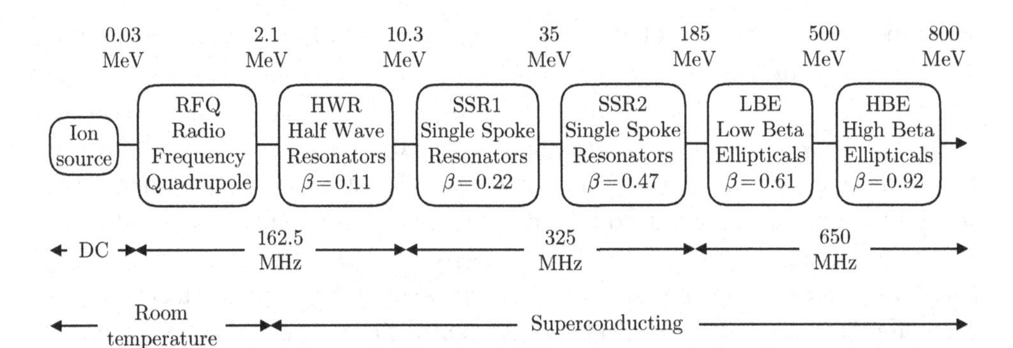

Figure 13.1 A typical technology map, for the PIP-II proton linac. Six technologies (RFQ, HWR, SSR1, SSR2, LBE and HBE) are constructed and tuned for different optimum values of β, the relativistic speed. Except for the single RFQ, each technology is repeated in many identical superconducting resonators of the same frequency, assembled into cryomodules. Elliptical RF cavities are also used in electron linacs (with $\beta = 1$).

single-pass linacs, in contrast to the transverse dynamics that usually dominate multi-pass circular rings.

13.1 Time Structures

The continuous beam leaving an ion source (usually) begins to be bunched in the first module of an RFQ, with a time spacing of $1/f_{RFQ}$, as illustrated in Figure 13.2. Typically about 10^6 identical bunches, spaced by a few nanoseconds, are combined into a single macropulse of order 1 millisecond long. The macropulse itself has a repetition rate in the range from 14 to 60 Hz, for the representative linacs in Table 13.1. Thus the beam power is

$$P = IVfT \tag{13.1}$$

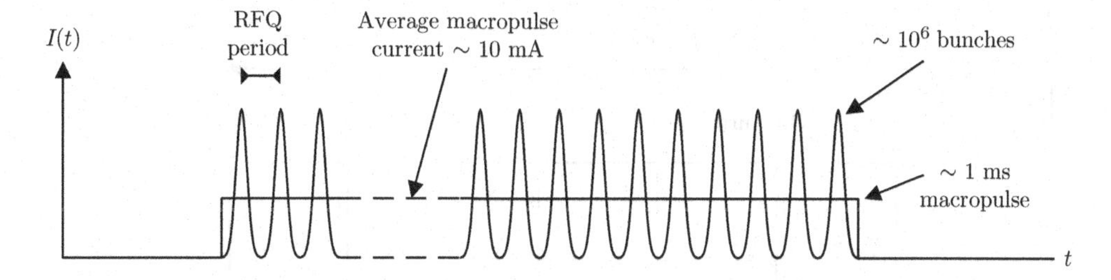

Figure 13.2 The bunches accelerated in a proton linac are typically separated by a few nanoseconds, forming a single macropulse that is of order a millisecond long. This structure is usually imposed by a Radio Frequency Quadrupole that creates of order a million bunches from a DC beam.

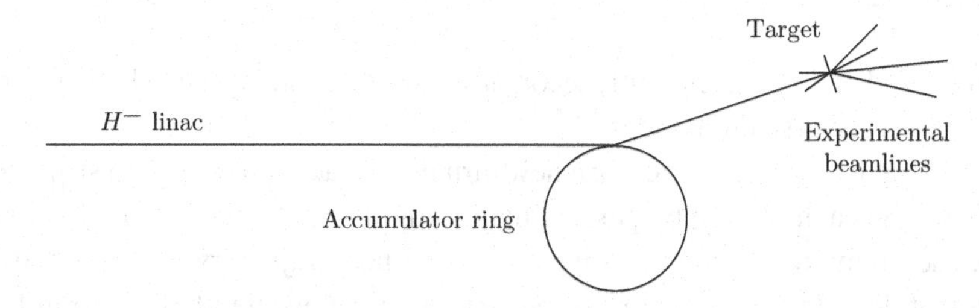

Figure 13.3 An H^- linac feeding an accumulator ring, in order to compress the beam in time. The single pulse of accumulated beam that is extracted from the ring is much shorter in time than the macropulse length (compressed into microseconds rather than milliseconds), but has much larger transverse emittances. The accumulator ring is not necessary in long pulse or continuous waveform operation, in which case protons are preferred.

where I is the (average) macropulse current, f is the macropulse repetition rate, T is the macropulse length and

$$V = \frac{W}{e} = \frac{m_p c^2}{e} (\gamma - 1) \tag{13.2}$$

is the kinetic energy of the proton or ion, measured in electron volts.

H^- ions are accelerated in some cases (such as PIP-II or the SNS), with two electrons attached to each proton. This enables a single macropulse to be accumulated in a compressor ring at the end of the linac, as illustrated in Figure 13.3. Short gaps in the bunch train chop the linac macropulse into (typically) about 1,000 segments, each fitting into one period of the accumulator, as illustrated in Figure 13.4. Successive turns are injected on top of each other in longitudinal phase space, but are 'painted' onto non-overlapping locations in horizontal and vertical phase space. When all the segments have been accumulated on successive turns from one macropulse, the time gap in the accumulated beam is used to ramp extraction magnets with minimal beam loss, delivering to a suitable target a beam pulse that

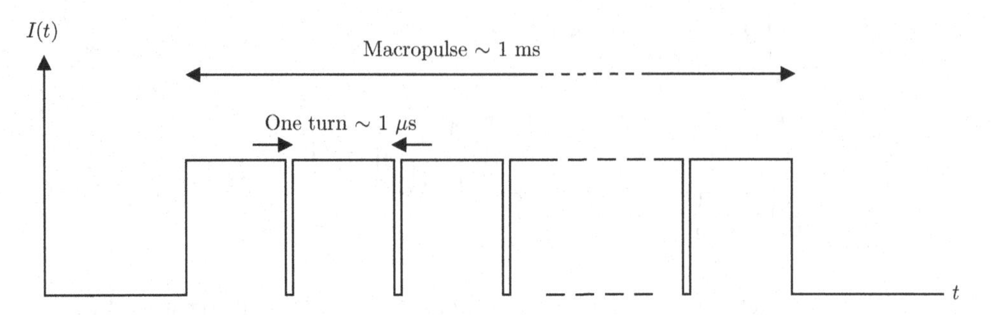

Figure 13.4 In an H^- macropulse, short gaps chop the macropulse into segments that can be accumulated and compressed in a storage ring, as shown in Figure 13.3.

is compressed in time from milliseconds to microseconds, but which is greatly enlarged in transverse emittances.

The crucial first step in injecting a new turn into an accumulator is to strip the two electrons from each H^- ion by passing the beam through a thin foil or a laser beam, so that each remaining proton is diverted onto a new trajectory. Accelerators such as ESS that do not need to compress the beam more conveniently use a simpler and more robust proton beam, while maintaining millisecond pulse structures that are appropriate (for example) for long-pulse spallation neutron sources. Accelerators that are capable of continuous waveform operation offer the highest beam powers, and are reasonably compatible with pulsed-beam operation at somewhat lower average currents.

The number of ions per bunch

$$N = 6.3 \times 10^6 \, \frac{I \,[\text{mA}]}{f_{RFQ} \,[\text{GHz}]} \tag{13.3}$$

is typically of order 10^8, in a bunch that tends to become shorter as it accelerates down the linac, so that the peak current increases with γ. It is convenient to increase the RF frequency at energies above about 10 MeV, because the volume of each RF structure scales like f^{-3}, and large transverse apertures are no longer needed. Higher frequencies save money. Figure 13.1 shows how the frequency doubles twice – from 162.5 MHz to 325 MHz to 650 MHz – in a typical proton linac. Despite frequency doubling, the bunches remain spaced at the RFQ period.

Considerable capital and operational costs are saved by decreasing the amount of stored energy that the high-power RF system provides to fill the cavities at the beginning of every macropulse. The low-level RF system – with controls, feedback and feedforward loops – must also stabilise before the first beam in a macropulse enters a linac. This is illustrated in Figure 13.5, which shows how the duty factor for the RF system

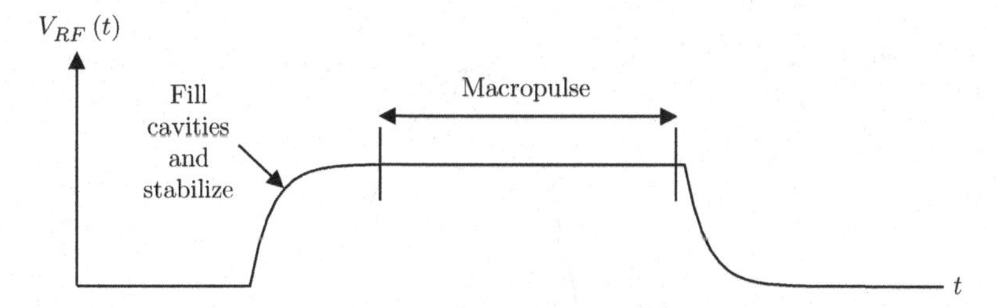

Figure 13.5 The RF voltage ramps up for around a millisecond before the beam arrives, to settle transients and to stabilise the low level RF controls.

$$D_{RF} \geq D_{macropulse} = fT \tag{13.4}$$

can be considerably larger than the macropulse duty factor.

Peak currents in proton linacs (with longer bunches and somewhat smaller RF frequencies) tend to be significantly less than in electron linacs. This is healthy, from the beam dynamics perspective. Not so healthy is the inescapable reality that the protons or ions are never fully relativistic.

13.2 Multi-Cell Synchronism

Section 7.4 shows that the maximum voltage V_A that a particle can acquire passing through a single-cell cavity resonating with a maximum gradient E_0 in a TM_{0n0} mode is

$$V_A = E_0 L \cdot T_1 \tag{13.5}$$

where the single-cell transit-time factor T_1 varies with β according to

$$T_1(\beta) = \frac{\sin(\omega L/2\beta c)}{\omega L/2\beta c} \tag{13.6}$$

for the particular case of a pill-box cavity of length L. To be more accurate, the on-axis accelerating field E_z generally varies with z at a fixed time (except for a pill-box cavity), and so

$$T_1(\beta) = \frac{\int_{-L/2}^{L/2} E_z \cos(\omega t(z))\, dz}{\int_{-L/2}^{L/2} E_z\, dz} \tag{13.7}$$

where

$$t(z) \approx \frac{z}{\beta c} \tag{13.8}$$

if β is reasonably constant across the cell. To a fair approximation, however, the pill-box expression of Equation 13.6 is valid for most RF cell structures in the

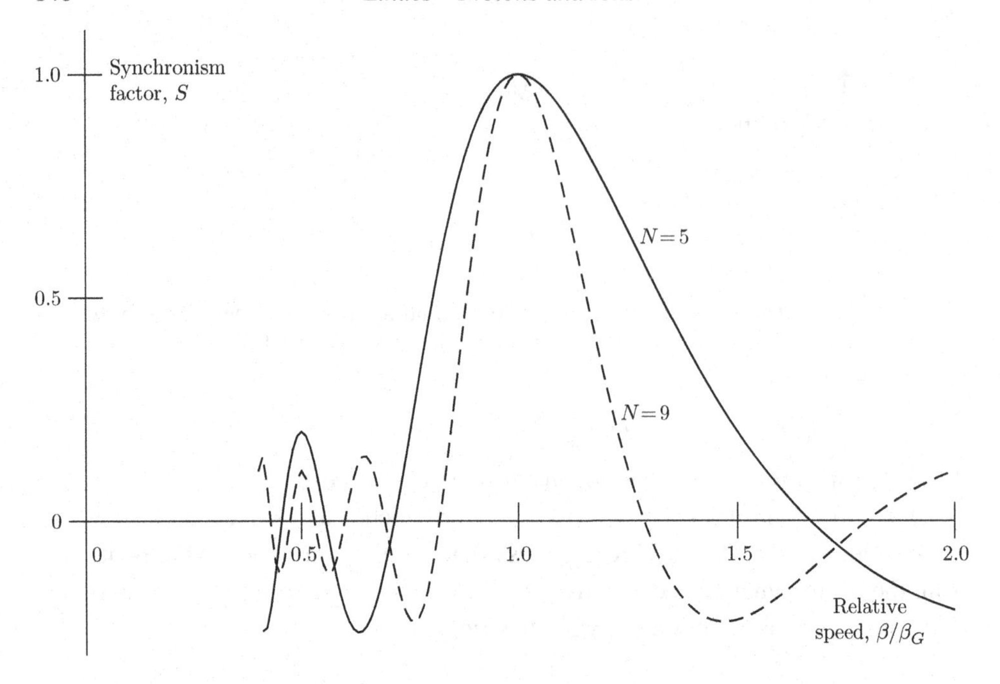

Figure 13.6 The multi-cell synchronism factor, $S(N, \beta/\beta_G)$. The useful dynamic range of $|\beta - \beta_G|$ scales like $1/N$, so proton linacs usually have fewer cells per cavity (e.g. $N = 5$) than electron linacs (e.g. $N = 9$).

proton linac menagerie, for example for the elliptical cavities sketched in Figure 7.5, or for half-wave resonators.

The maximum accelerating voltage is readily increased by combining multiple cells into a single cavity, as shown for five cells in Figure 7.5. If N identical cells have a phase advance of π between them (at fixed time), then

$$V_A = NE_0L \cdot T_1S \tag{13.9}$$

where the synchronism factor shown in Figure 13.6

$$\begin{aligned} S(\beta) &= (1/N)\left[1 + \sum_{m=1}^{(N-1)/2} (-1)^m \, 2\cos(m\pi\beta_G/\beta)\right] \qquad N \text{ odd} \\ &= \quad (2/N) \sum_{m=0}^{N/2-1} (-1)^m \sin\left((m + \tfrac{1}{2})\pi\beta_G/\beta\right) \qquad N \text{ even} \end{aligned} \tag{13.10}$$

has a maximum value of $S = 1$ at $\beta = \beta_G$, when the relativistic speed of the particle is perfectly matched to the geometric beta of the cavity [57]. The cavity becomes less efficient – accelerates less – the further that β is from the ideal value, with a useful dynamic range

$$|\beta - \beta_G| < \frac{1}{N} \tag{13.11}$$

determined by N. This limited dynamic range is not an impediment in electron linacs, when $\beta \approx 1$ even at the lowest energies, but it limits the usefulness of each cavity geometry in hadron linacs, especially in non-relativistic heavy ion linacs that are intended to accelerate any of a broad set of ion species. Proton linacs usually have no more than $N = 5$ cells per cavity, while there are typically $N = 9$ cells per cavity in electron linacs. Proton linacs have multiple families of cavities, each with a different value of β_G.

Non-relativistic speeds have some advantages. For example, time-of-flight measurements of the difference in bunch arrival time from one module to the next lead directly to beam energy measurements. This technique is more accurate and useful at lower energies, for example in tuning Drift Tube Linacs.

13.3 Linear Motion

A thin cavity approximation is reasonable in many cases: at larger values of β, and when β does not change much across the cavity. All the energy is then gained from an impulse at the electrical centre of thin cavity n, so

$$\Delta W_n = qV_{A,n}(\beta_n)\cos(\phi_n) \tag{13.12}$$

where q is the charge of the proton or ion, and the maximum voltage gain $V_{A,n} \sim T_1 S$ depends on W_n (via β_n) through Equation 13.9. The RF phase ϕ_n depends on the time of arrival at the electrical centre, as illustrated in Figure 13.7 – earlier particles are accelerated less.

Unlike a circular accelerator there is no natural synchronous phase in a proton linac, and so small deviations in ϕ and W are referred instead to an idealised reference particle that has energy and phase $(\phi_{r,n}, W_{r,n})$ in cavity n, so that

$$\begin{aligned} \delta\phi_n &= \phi_n - \phi_{r,n} \\ \delta W_n &= W_n - W_{r,n} \end{aligned} \tag{13.13}$$

A particle with positive δW_n goes faster and tends to arrive early, thereby receiving less acceleration, slowing down relative to the reference particle, and being restored towards the reference phase. It remains bunched with the reference particle, performing a small number of longitudinal oscillations, so long as it is launched close enough to the origin in $(\delta\phi, \delta W)$ phase space.

The change in phase advance from cavity n to $n + 1$ is

$$\delta\phi_{n+1} = \frac{2\pi(s_{n+1} - s_n)}{\lambda_{RF}}\left(\frac{1}{\beta_n} - \frac{1}{\beta_{r,n}}\right) + \delta\phi_n \tag{13.14}$$

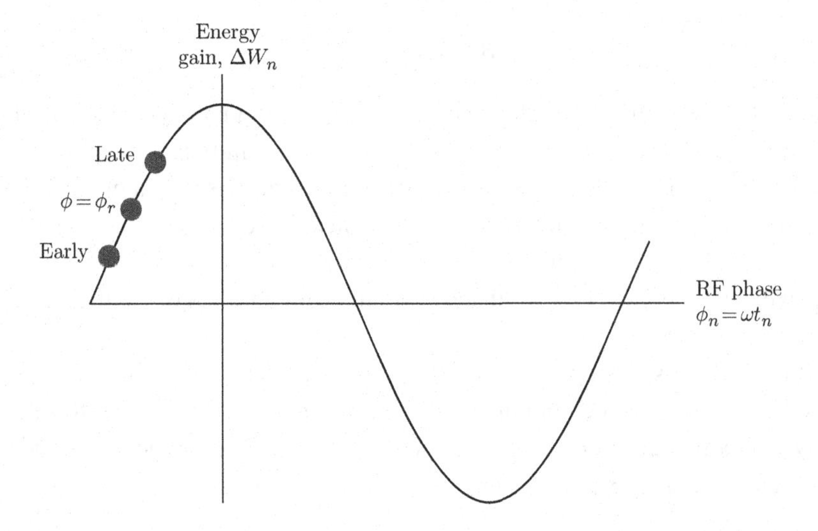

Figure 13.7 Energy gain and reference phase ϕ_r at the electrical centre of a thin cavity. Early and late particles receive a relative restoring force that keeps them bunched with the reference particle. See Figure 11.4 for a comparison with the synchronous particle in a circular accelerator.

where $s_{n+1} - s_n$ is the longitudinal distance between electrical centres. In linearised matrix form

$$\begin{pmatrix} \delta\phi \\ \delta W \end{pmatrix}_{n+1} = M_{L-DRIFT} \begin{pmatrix} \delta\phi \\ \delta W \end{pmatrix}_n \tag{13.15}$$

where the longitudinal drift matrix

$$M_{L-DRIFT} = \begin{pmatrix} 1 & L_e \\ 0 & 1 \end{pmatrix} \tag{13.16}$$

is analogous to the transverse drift matrix described in Equation 2.24. The effective length of this drift is negative

$$L_e = -\frac{1}{mc^2} \frac{1}{\beta_r^2 \gamma_r^3} \frac{2\pi (s_{n+1} - s_n)}{\lambda_{RF}} \tag{13.17}$$

since a higher energy particle decreases its phase. Electrons do not slip in phase, to all practical purposes, thanks to the strong dependence of L_e on γ_r^{-3}, and so may be accelerated 'on-crest' with $\phi_r = 0$ in every cavity.

In contrast, the linearised matrix for a thin cavity is unfortunately not simply analogous to the thin quadrupole matrix of Equation 2.26, since the energy gain

$$\delta W_n = q \left(V_{A,n} \cos\phi_n - V_{A,r,n} \cos\phi_{r,n} \right) + \delta W_{n-1} \tag{13.18}$$

depends not only on $\delta\phi$, but also on δW, through the variation of $V_{A,n}$ and $V_{A,r,n}$ with δW via $\delta\beta$. Worse, the energy gain is nonlinear in both $\delta\phi$ and δW, even for particles launched modestly close to the phase space origin, not only because of the cosine dependence in Equation 13.18, but also because T_1 and S are nonlinear in δW.

Radial Defocusing

Proton linacs inherently defocus in both transverse planes through radial defocusing in RF cavities, even when weak beams do not suffer from repellent space charge forces. This is mainly rooted in the need to keep the beam longitudinally bunched, with negative reference phases, although two other cavity effects also contribute at lower values of β: field dependence on radial displacement, and differences in particle speed between the two cell ends [57].

Consider the perturbation that the beampipe holes make to the TM_{0n0} pill-box fields described in Equation 7.12, and as sketched in Figures 7.4 and 13.8, under the assumption that

$$E_z(r,z,t) = E_0(z)\cos(\omega t + \phi) \tag{13.19}$$

close to the axis. Maxwell's equations require that, close to the axis of a general rotationally symmetric cell [57]

$$E_r = -\frac{r}{2}\frac{\partial E_z}{\partial z} \qquad B_\theta = \frac{r}{2c^2}\frac{\partial E_z}{\partial t} \tag{13.20}$$

so that the net radial momentum impulse across the cell is

$$\Delta p_r = q\int_{-\infty}^{\infty}(E_r - \beta c B_\theta)\frac{dz}{\beta c} = -\frac{q}{2}\int_{-\infty}^{\infty} r\left(\frac{\partial E_z}{\partial z} + \frac{\beta}{c}\frac{\partial E_z}{\partial t}\right)\frac{dz}{\beta c} \tag{13.21}$$

where the radial momentum is

$$p_r = m\,\beta\gamma c\, r' \tag{13.22}$$

This expression is simplified by using the two identities

$$\frac{dE_z}{dz} = \frac{\partial E_z}{\partial z} + \frac{1}{\beta c}\frac{\partial E_z}{\partial t} \tag{13.23}$$

and

$$\int_{-\infty}^{\infty}\frac{dE_z}{dz}\,dz = 0 \tag{13.24}$$

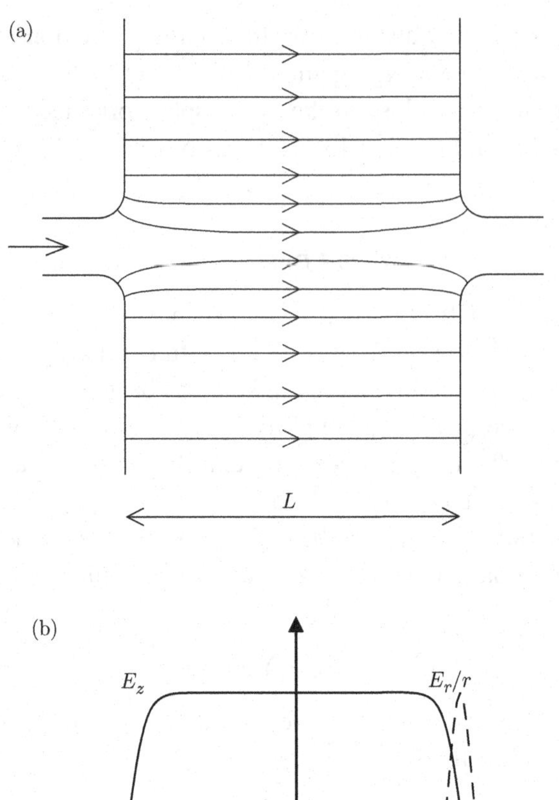

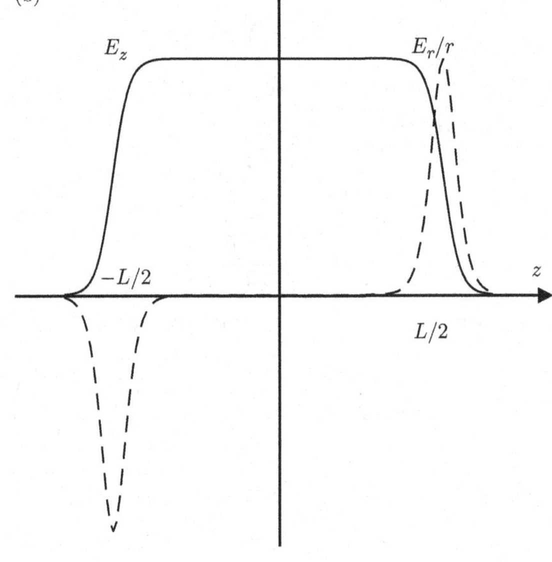

Figure 13.8 Unbalanced radial impulses at the ends of a rotationally symmetric cavity. Radial electric fields inevitably occur where the beampipes intrude at the cavity ends in (a). They are linear in r for small r, defocusing and focusing at entry and exit respectively, as shown in (b). The defocusing effect dominates, because the exiting beam is more rigid.

so that

$$\Delta p_r = \frac{qr}{2\beta c}\int_{-\infty}^{\infty}\left(\frac{1}{\beta c}-\frac{\beta}{c}\right)\frac{\partial E_z}{\partial t}\,dz \tag{13.25}$$

$$= \frac{-qr\omega}{2\beta^2\gamma^2c^2}\int_{-\infty}^{\infty} E_0(z)\,\sin(\omega t+\phi)\,dz$$

$$= \frac{-qr\omega}{2\beta^2\gamma^2c^2}\sin(\phi)\int_{-\infty}^{\infty} E_0(z)\,\cos\left(\frac{2\pi z}{\beta\lambda}\right)dz$$

where it is assumed that r and β are approximately constant across the cell, and $t = 0$ when $z = 0$.

Finally, the net radial focusing for a reference particle with $\phi = \phi_r$ is

$$\Delta r' = \frac{\Delta p_r}{p} = -\frac{\pi q E_0 T_1 L}{mc^2\beta^3\gamma^3\lambda}\cdot\sin(\phi_r)\cdot r \tag{13.26}$$

where T_1 is the single-cell transit-time factor. Crucially, the reference phase ϕ_r must be negative in proton linacs for longitudinal focusing, guaranteeing radial defocusing. Longitudinal and transverse focusing are incompatible, using RF structures alone, in this approximation! The strong dependence on $(\beta\gamma)^{-3}$ shows a rapid weakening of this effect at higher proton energies. It also shows why electron linacs only suffer from radial defocusing at MeV-scale energies, and lower.

Transverse Focusing

Radio Frequency Quadrupoles provide transverse focusing at the lowest proton energies, up to a few MeV, while also accelerating and bunching the beam longitudinally. This is discussed in Section 13.4.

Solenoids are effective in overcoming radial defocusing at intermediate energies. A particle entering a solenoid with horizontal displacement $x = r$ parallel to the axis with initial angle $r' = 0$ follows a trajectory

$$\begin{pmatrix} x(s) \\ y(s) \end{pmatrix} = r\begin{pmatrix} \cos^2(Ks) \\ -\sin(Ks)\cos(Ks) \end{pmatrix} = \frac{r}{2}\begin{pmatrix} 1+\cos(2Ks) \\ -\sin(2Ks) \end{pmatrix} \tag{13.27}$$

according to Equation A.22, where the geometric strength

$$K = \text{sgn(q)}\,\frac{B}{2(B\rho)} \tag{13.28}$$

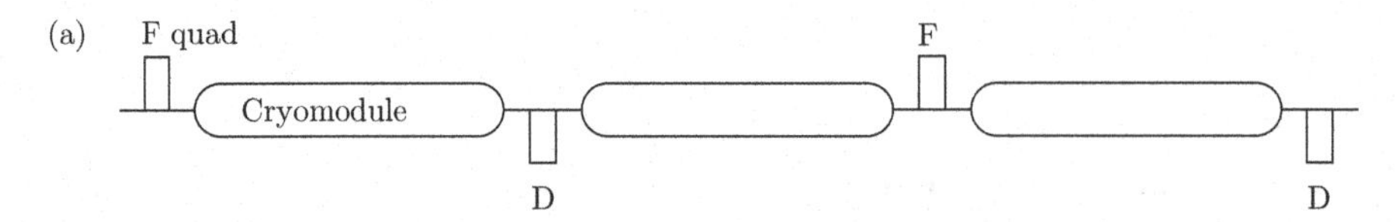

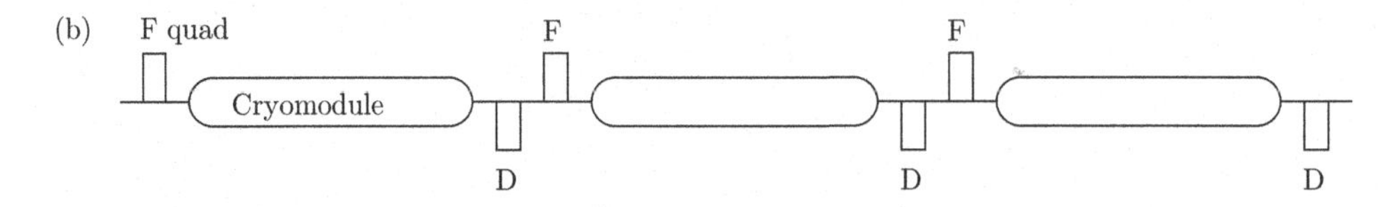

Figure 13.9 Singlet (a) and doublet (b) focusing, with normal-conducting quadrupoles placed between superconducting cryomodules. Doublet focusing increases the linac length slightly, but approximately halves the length of the optical period, and delivers rounder beams.

is linear in the solenoid field strength B. If the solenoid is thin, with $KL \ll 1$, then the particle receives a radial angular kick of

$$\Delta r' \approx -\frac{K}{2}\sin(2KL)\,r \approx -\left(\frac{B}{2(B\rho)}\right)^2 L\,r \tag{13.29}$$

showing that, while a solenoid focuses in both planes simultaneously, its strength falls off quadratically with the rigidity $(B\rho)$, and with the longitudinal momentum.

Quadrupoles weaken only linearly with rigidity. They can be directly incorporated *inside* the drift tubes of a Drift Tube Linac module, as permanent magnets or electromagnets. At higher energies quadrupoles are often located between RF modules, as sketched in Figure 13.9 for superconducting cryomodules that may contain elliptical cavities, or spoke resonators, etc. It is operationally convenient and conventional to place normal-conducting magnets (and other instrumentation) in between segmented cryomodules in this way, although the many warm-to-cold transitions increase the static thermal load, and lengthen the linac. In contrast, the highest-energy electron linacs have continuous cryostats, containing superconducting quadrupoles.

If two short quadrupoles in an FD (or DF) doublet have equal and opposite focal lengths $\pm f$, and are separated by a short distance L, then they have a net focal strength of

$$\frac{1}{f_{net}} \approx -\frac{L}{f^2} \tag{13.30}$$

in both horizontal and vertical planes. Because of this, the horizontal and vertical β-functions (and beam sizes) inside cryomodules separated by doublet quadrupoles are much more equal than in cryomodules separated by singlet quadrupoles. Also, the periodic cell length with doublet focusing is approximately half of the cell length achieved with singlet focusing. Both of these effects – rounder beams and shorter periodicity – help doublet focusing to ameliorate space charge effects. Doublet focusing is more common than singlet focusing in proton linacs. However, doublet focusing requires more total linac length, since Equation 13.30 motivates the natural tendency to increase L, to enable weaker quadrupoles. Both cryomodule segmentation and doublet focusing increase the length of proton linacs.

The calculation of radial defocusing, above, found that the defocusing at one end of a cavity is stronger than the focusing at the other end, and simply added the two together to generate Equation 13.26. A more rigorous calculation recognises the doublet nature of radial defocusing, with unequal $1/f_D$ and $1/f_F$ strengths at the two – separated – ends.

13.4 Radio Frequency Quadrupoles

The 1969 invention of the Radio Frequency Quadrupole was a major step forward in enabling high current proton and ion linacs [23]. An RFQ is essentially an electrostatic device, since RF electric fields dominate RF magnetic fields when the proton speed is small. Not only does an RFQ accelerate protons – typically from about 100 keV to a few MeV ($0.01 \lesssim \beta \lesssim 0.06$) – but it also bunches them longitudinally, and focuses transversely. The focusing is so strong that currents up to about 100 mA can be accelerated without significant beam disruption or blowup due to defocusing space charge forces.

A cross-section of the four conducting vanes in a representative RFQ is shown on the left of Figure 13.10, at a location $z = 0$ where the horizontal vanes touch a radius of a, and the vertical vanes reach a larger radius of ma, where $m > 1$ is called the modulation parameter. The vane separation is modulated along the length of the RFQ with a period L, so that the horizontal vanes touch a maximum radius of ma at $z = L/2$, where the vertical vanes simultaneously reach a minimum of a.

The voltages on the four vanes are excited at an RF frequency with a quadrupole symmetry

$$V = \pm\frac{V_0}{2}\cos(\omega t) \tag{13.31}$$

where the positive sign applies to both horizontal vanes, as shown in Figure 13.10 for $t = 0$. Since the horizontal vanes are closer than the vertical vanes at $z = 0$, the net potential is positive on-axis. Similarly, the on-axis potential is negative at $z = L/2$, and so there is an accelerating field over the range $0 < z < L/2$ when $t = 0$. At the same time a proton experiences a horizontally focusing electric field, and a vertically defocusing field, for all values of z.

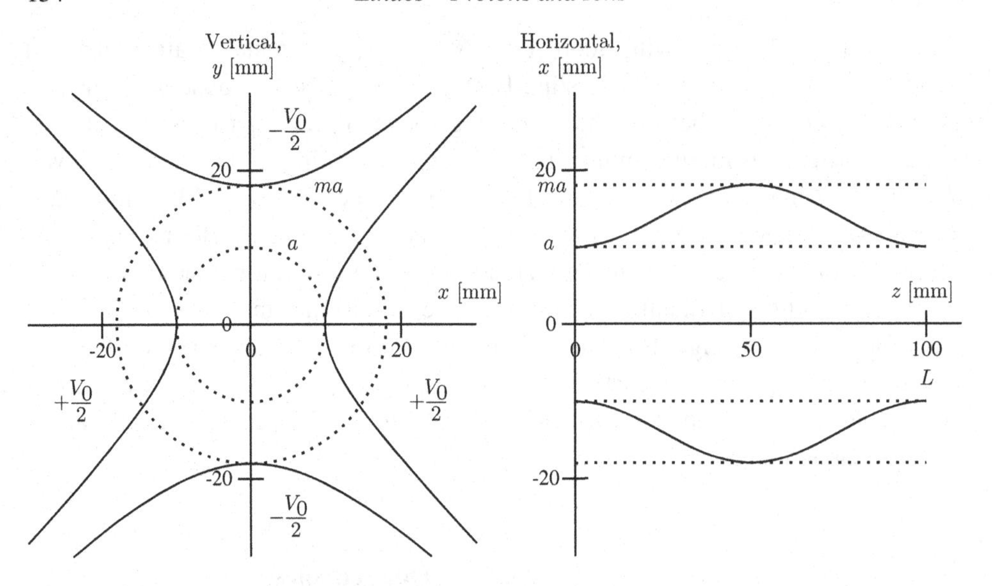

Figure 13.10 Cross-sectional and top views of a typical RFQ vane geometry, with minimum vane radius $a = 10$ mm, $m = 1.8$ and $L = 100$ mm. The vertical vane separation also oscillates with longitude z, between radii of a and ma, but out of phase with the horizontal.

In general the Kapchinsky–Teplyakov potential with quadrupole symmetry is

$$V = \sum_{j=0}^{\infty} A_j r^{2(2j+1)} \cos\left(2(2j+1)\theta\right) + \sum_{p=1}^{\infty}\sum_{j=0}^{\infty} A_{pj} I_{2j}(pkr)\cos(2j\theta)\cos(pkz) \tag{13.32}$$

where $k = 2\pi/L$ and I is a modified Bessel function [56, 57]. Optimised vane shapes (not just simple hyperbolae) eliminate all but two of the terms, so

$$V = A_0 r^2 \cos(2\theta) + A_{10} I_0(kr)\cos(kz) \tag{13.33}$$

The electric fields obtained by taking the gradient of V are

$$E_x = -X\left(V_0/a^2\right)x - A\left(\frac{kV_0 I_1(kr)}{2r}\right)\cos(kz)\,x \tag{13.34}$$

$$E_y = X\left(V_0/a^2\right)y - A\left(\frac{kV_0 I_1(kr)}{2r}\right)\cos(kz)\,y$$

$$E_z = A\left(\frac{kV_0 I_0(kr)}{2}\right)\sin(kz)$$

where the dimensionless acceleration and focusing efficiencies are

$$A = \frac{m^2 - 1}{m^2 I_0(ka) + I_0(kma)} \tag{13.35}$$
$$X = \frac{I_0(ka) + I_0(kma)}{m^2 I_0(ka) + I_0(kma)}$$

respectively. For modest arguments $\nu \lesssim 1$

$$I_0(\nu) \approx 1 + \frac{\nu^2}{4} \tag{13.36}$$
$$I_1(\nu) \approx \frac{\nu}{2}$$

and so if

$$kma = 2\pi \frac{ma}{L} \lesssim 1 \tag{13.37}$$

then

$$A + X \approx 1 \tag{13.38}$$

as confirmed by Figure 13.11 for the vane geometry of Figure 13.10. The value of A increases as m is increased (for fixed values of a and L), delivering more

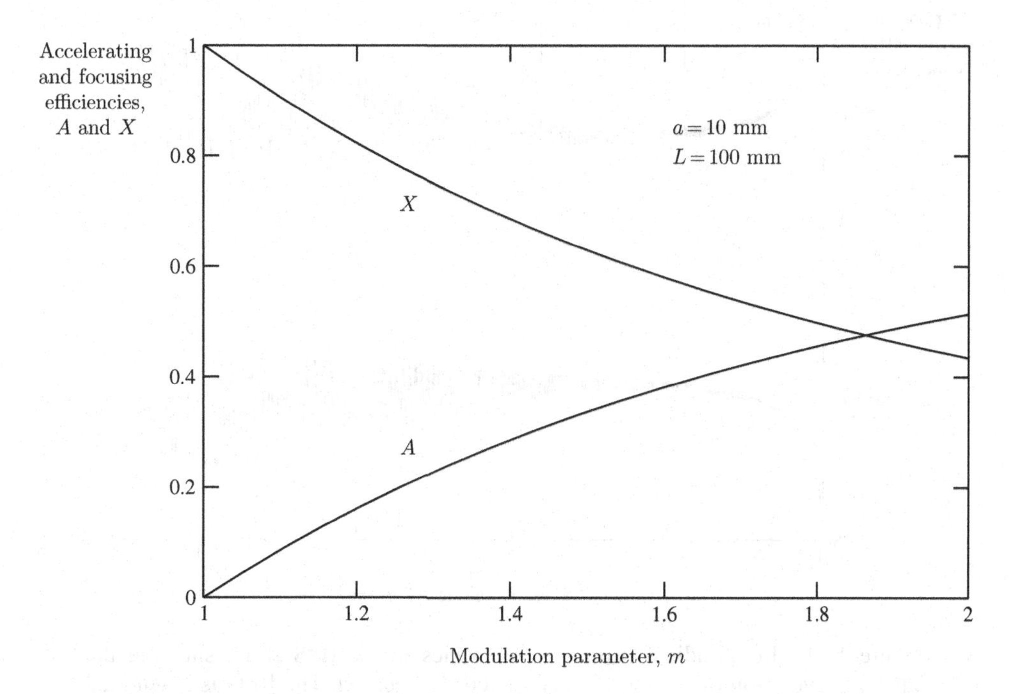

Figure 13.11 RFQ acceleration and focusing efficiencies, A and X, versus modulation parameter m, for the typical vane geometry in Figure 13.10.

acceleration at the expense of decreasing X and weakening the transverse focusing. The linearised fields for small x and y are

$$E_x \approx V_0 \left(-\frac{X}{a^2} - \frac{\pi^2 A}{L^2} \cos(kz) \right) x \tag{13.39}$$
$$E_y \approx V_0 \left(\frac{X}{a^2} - \frac{\pi^2 A}{L^2} \cos(kz) \right) y$$
$$E_z \approx V_0 \frac{\pi A}{L} \sin(kz)$$

showing that the geometric ratio a/L is vital in trading off A against X. Each field is also modulated in time by $\cos(\omega t)$.

Continuous acceleration of a reference particle with speed β_r requires that the polarity of the field changes twice per period L, so that

$$L = \beta_r \lambda \tag{13.40}$$

where λ is the free-space wavelength corresponding to the resonant frequency ω at which the RFQ cavity is excited. L increases along the length of the RFQ, to synchronise with the reference particle as β_r increases. The other two control

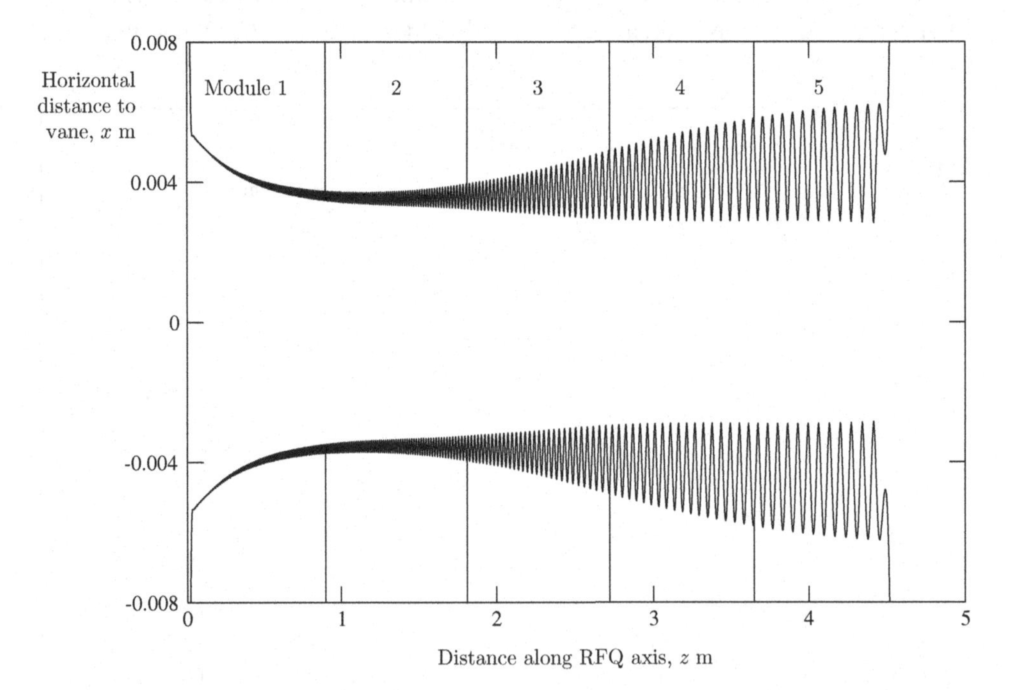

Figure 13.12 Longitudinal profile of the vanes in the ESS RFQ, showing the adiabatic evolution of the control parameters L, a and m. The RFQ is assembled from five modules to form a single resonant cavity that is excited at 352 MHz. (Courtesy of A. Ponton.)

parameters a and m also evolve with z in an optimised design, as shown for the 4.6 m length of the ESS RFQ in Figure 13.12. Continuous beam entering the RFQ is immediately subject to strong transverse focusing, with values of m that are initially only just larger than 1, causing gradual beam bunching. Acceleration is enhanced by adiabatically increasing m to a final value of about 2 when bunching is well established.

RFQs have significant advantages over the DC electrostatic systems that they superseded: continuous transverse electrostatic focusing is stronger than occasional magnetic focusing at slow speeds, efficient adiabatic bunching begins almost immediately out of the ion source and the control parameters (L, a, m) are continuously optimised by machining the vanes. But RFQs also have their disadvantages: high tolerances must be maintained in machining complex structures that are many meters long, thermo-mechanical stability is a challenge with higher power beams and only one charge-to-mass ratio can be accelerated in a particular RFQ. Nonetheless, 10 MW-class proton linacs are only practical thanks to RFQ technology.

13.5 Beam Losses and Haloes

Unintended beam losses must be reduced to less than about 1 W/m at general locations in the linac tunnel, in order to reduce radio-activation of components, and to enable immediate hands-on maintenance of equipment. Only about 10^{-4} of the total current of a 10 MW proton beam can be lost during its transmission along a linac (except under special circumstances: at very low energies, or at well-shielded collimators). This motivates the theoretical and experimental study of the creation and evolution of very weak beam haloes in the tails of the 3-D beam distribution. Such studies are incomplete and ongoing at the time of writing.

Intrabeam stripping can be a dominant mechanism for beam losses in well-tuned H^- linacs, not far below the 1 W/m level. Coulomb scattering between two co-moving H^- ions causes one to lose one or two of its electrons, and then be lost [28]. This mechanism is absent in proton linacs. However, if multi-turn accumulation and compression is necessary, then H^- beams are inevitable.

Longitudinal motion in typical proton linacs is strongly nonlinear, while transverse motion is often essentially linear. This is in stark contrast to typical circular accelerators, where the converse is true. Nonetheless, experience shows that operational cavity strengths can deviate from design values at the level of 10% or more, without undue influence on beam loss rates.

Protons that somehow *do* fall out of a longitudinal RF bucket are ballistic at something approaching the speed of light. They might reasonably be expected to reach the linac target in any case, without hitting the beampipe walls. However, such protons can become over-focused transversely by quadrupoles with strengths appropriate to much higher energies. Transverse linear instability causes exponential amplitude growth, driving protons or ions into the walls.

Space charge is the elephant in the room. A thorough discussion of proton linac design, or quadrupole or RFQ strengths, or beam losses, must include space charge. For example, it may not be possible to reduce quadrupole strengths (to ameliorate over-focusing) because of the need to combat space charge distortions of the linear optics. However, an appropriate discussion of space charge is not possible within the confines of this chapter, which concentrates on weak beam single particle dynamics.

Exercises

13.1 Derive Equation 13.17.

13.2 Expand Equation 13.18 to derive the linearised matrix for a thin cavity.

13.3 Prove Equation 13.30.

14

Linacs – Electrons

Electrons in even modest-energy linacs are highly relativistic, in contrast to the protons in the linacs described in Chapter 13. Electrons with an energy of 10 MeV are already within about 10^{-3} of the speed of light, so the ultra-relativistic approximation ($\beta \approx 1, \gamma \gg 1$) is reasonable throughout this chapter.

All accelerating structures can be identical in an electron linac, with constant length and equal spacing after a short low-energy injector. Accelerating structures may be single-cell or multi-cell standing-wave RF cavities (as in proton and ion linacs), or may be loaded RF waveguides that carry travelling waves with a phase velocity $v_P = \beta c$ that matches the electron velocity.

As many as $N = 9$ cells are typically used in multi-cell cavities, limited by tuning and higher-order mode (HOM) performance rather than by the synchronism factor shown in Figure 13.6. The design gradient for the 1.3 GHz superconducting cavities originally developed for the ILC is 30 MV/m, limited by field emission and peak surface fields. Very similar cavities are used in linacs such as XFEL and LCLS-II, as illustrated by the technology map in Figure 14.1. Loaded RF waveguide length is typically limited by wakefields, resistive wall losses and ultimately by surface electric field breakdown. Short-pulse waveguides operating at 12 GHz for CLIC have achieved gradients of up to 100 MV/m, but with low repetition rates.

Linac RF technologies are discussed in detail elsewhere [57]. This chapter focuses instead on dynamical issues, and on an abbreviated survey of electron linac configurations.

14.1 Longitudinal and Transverse Focusing

When an electron moves from one accelerating structure to the next, small relative deviations in phase and energy evolve like

$$\begin{pmatrix} \delta\phi \\ \delta W \end{pmatrix}_{n+1} = \begin{pmatrix} 1 & L_e \\ 0 & 1 \end{pmatrix} \begin{pmatrix} \delta\phi \\ \delta W \end{pmatrix}_n \tag{14.1}$$

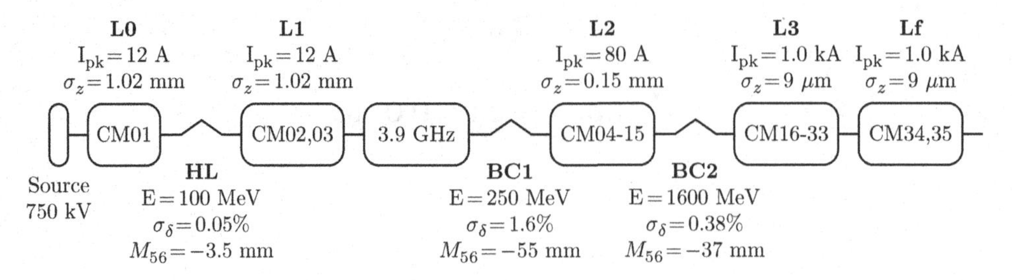

Figure 14.1 The technology map of the LCLS-II electron linac, for comparison with the proton linac technology map of Figure 13.1. Except for the 750 kV gun and the third-harmonic 3.9 GHz harmonic lineariser (HL), all accelerating structures are 1.3 GHz eight-cell superconducting elliptical cavities in 35 identical CM cryomodules [50]. BC1 and BC2 are bunch compressors.

in the linearised matrix form introduced in Section 13.3. The effective drift length is negative

$$L_e = -\frac{1}{mc^2}\frac{1}{\beta^2\gamma^3}\frac{2\pi(s_{n+1}-s_n)}{\lambda_{RF}} \tag{14.2}$$

since the phase of a higher energy particle decreases. Longitudinal oscillations stop in the limit $\gamma^3 \gg 1$ that is valid for electrons, and the longitudinal distribution is essentially frozen with respect to the reference particle. Electrons do not (necessarily) need longitudinal focusing! Often they can accelerate optimally on the crest of the RF wave, gaining maximum energy per structure.

The bunch length in the lab frame compresses naturally as the bunch is accelerated, due solely to relativistic effects. Off-crest acceleration is performed if it is necessary to manipulate the electron bunch length, momentum spread or peak current, for example during bunch compression (Section 14.3) or to stabilise against beam breakup instabilities (Section 14.5).

A reference electron with phase $\phi = \phi_r$ passing through an accelerating structure is radially defocused

$$\Delta r' = -\frac{\pi q E_0 T_1 L}{mc^2\beta^3\gamma^3\lambda}\cdot\sin(\phi_r)\cdot r \tag{14.3}$$

according to Equation 13.26. Radial defocusing also shows a strong inverse dependence on γ^3, making it, too, negligible. Magnetic focusing is only required to establish control of the transverse optics, and to provide some chromaticity to protect against instabilities such as beam breakup (Section 14.5). Long cell FODO optics are adequate, with a phase advance per cell in the range 60° to 120°, and with quadrupoles placed between successive acceleration modules. Quadrupole gradients must be properly scaled for the local beam energy and rigidity, in order to avoid over-focusing.

14.2 RF Capture

The initial capture of relatively low-energy electrons is an interesting and practical beam dynamics problem. There is a lower limit on the peak strength E_0 of a co-propagating travelling wave that must be exceeded if an electron of initial speed $\beta_0 c$ is to be captured and accelerated through the remainder of the linac [27].

Consider a travelling wave that propagates at the speed of light, delivering an axial accelerating field

$$E_z = E_0 \sin\phi \tag{14.4}$$

where ϕ is the phase angle between the electron and the field. The distance l that the electron lags behind the accelerating wave increases with time, and so the phase angle evolves according to

$$\frac{d\phi}{dt} = \frac{2\pi}{\lambda_{\rm rf}}\frac{dl}{dt} = \frac{2\pi c}{\lambda_{\rm rf}}(1-\beta) \tag{14.5}$$

where $\lambda_{\rm rf}$ is the wavelength of the RF field, and βc is the local speed of the electron.

The electron also accelerates, following the equation of motion

$$\frac{dp}{dt} = \frac{d}{dt}(mc\beta\gamma) = mc\frac{d}{dt}\left[\frac{\beta}{\sqrt{1-\beta^2}}\right] = eE_0\sin\phi \tag{14.6}$$

Define χ through

$$\cos\chi \equiv \beta \tag{14.7}$$

so that

$$\frac{d\chi}{dt} = -\frac{eE_0}{mc}\sin\phi\,\sin^2\chi \tag{14.8}$$

Then, using

$$\frac{d\phi}{d\chi} = \frac{d\phi/dt}{d\chi/dt} \tag{14.9}$$

and Equation 14.5, the equation of motion for ϕ versus χ becomes

$$-\sin\phi\,\frac{d\phi}{d\chi} = \frac{2\pi}{\lambda_{\rm rf}}\frac{mc^2}{eE_0}\frac{(1-\cos\chi)}{\sin^2\chi} \tag{14.10}$$

where terms in ϕ and χ are separated on the left and right hand sides. Integrating Equation 14.10 from time t_0 to t_1

$$\cos\phi_0 - \cos\phi_1 = \frac{2\pi}{\lambda_{\rm rf}}\frac{mc^2}{eE_0}\left[\tan\frac{\chi_0}{2} - \tan\frac{\chi_1}{2}\right]. \tag{14.11}$$

and using the identity

$$\tan\frac{\chi}{2} = \sqrt{\frac{1-\beta}{1+\beta}} \tag{14.12}$$

finally gives

$$\cos\phi_0 - \cos\phi_1 = \frac{2\pi}{\lambda_{\rm rf}}\frac{mc^2}{eE_0}\left[\sqrt{\frac{1-\beta_0}{1+\beta_0}} - \sqrt{\frac{1-\beta_1}{1+\beta_1}}\right] \tag{14.13}$$

The left hand side of Equation 14.13 must be less than 2 if ϕ_1 is to remain real while β_1 approaches 1.

The RF capture condition on the peak accelerating electric field E_0 is therefore

$$E_0 \geq \frac{\pi\, mc^2}{\lambda_{\rm rf} e}\left[\sqrt{\frac{1-\beta_0}{1+\beta_0}} - \sqrt{\frac{1-\beta_1}{1+\beta_1}}\right] \tag{14.14}$$

In most practical situations $\phi_1 \to \pi/2$ as $\beta_1 \to 1$; electrons gather near the crest of the RF wave. For $\lambda_{\rm rf} = 0.2$ m and $\beta_0 = 0.6$, $E_0 \geq 4$ MV/m. This is a reasonable and achievable peak axial electric field, even for normal conducting RF cavities.

14.3 Bunch Compression

Very short electron bunches with very high peak currents are required in applications such as free electron lasers and linear colliders. Bunch compressors – beamline sections that longitudinally focus the electron beam – can also be applied in reverse, to lengthen electron bunches and reduce peak currents, for example to avoid effects such as coherent synchrotron radiation. In practice, the bunch may be compressed more than once. For example, two bunch compressors are used in LCLS-II to shorten electron bunches from $\sigma_z = 1.02$ mm to 9.0 μm, as illustrated in Figure 14.1.

A bunch compressor has two parts: an energy modulator that provides longitudinal focusing, and a dispersive section that acts like a longitudinal drift. The modulator may be just a single RF cavity, while the most common dispersive section is a chicane with four dipoles (and no quadrupoles). Figure 14.2 shows a simple chicane that shortens the path length of higher momentum particles, relative to lower momentum particles.

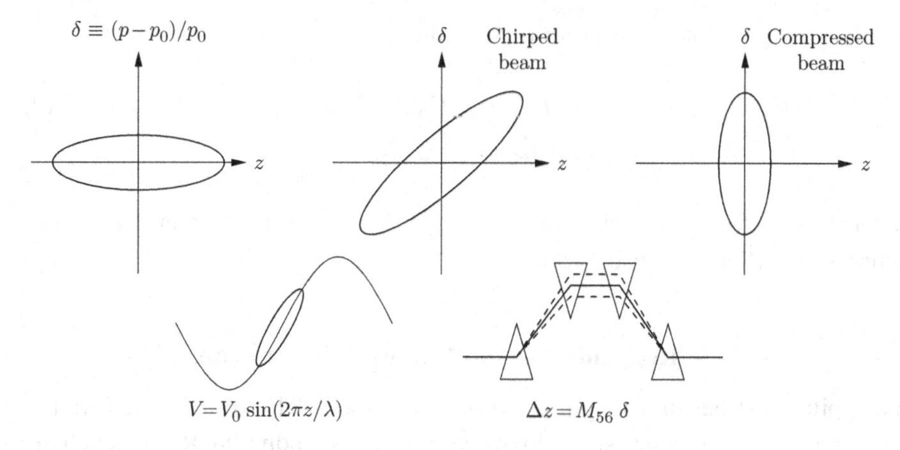

Figure 14.2 The longitudinal phase space evolution of a bunch undergoing bunch compression. First, the bunch passes through an energy modulator that tilts the phase space ellipse and imprints a chirp on the beam. Then the bunch passes through a dispersive section – in this case a four-dipole chicane – that results in an erect ellipse that is shorter, but higher, than when it entered the bunch compressor.

In a linear approximation of a thin RF cavity, the relative momentum offset δ changes but the longitudinal offset z does not, so

$$\begin{pmatrix} z_1 \\ \delta_1 \end{pmatrix} = \begin{pmatrix} 1 & 0 \\ M_{65} & M_{66} \end{pmatrix} \begin{pmatrix} z_0 \\ \delta_0 \end{pmatrix} \tag{14.15}$$

where

$$M_{65} = \left(\frac{eV_{\rm rf}}{U_0}\right) \frac{2\pi}{\lambda_{rf}} \sin \phi_{\rm rf} \tag{14.16}$$

$$M_{66} = 1 - \left(\frac{eV_{\rm rf}}{U_0}\right) \cos \phi_{\rm rf} \tag{14.17}$$

and the cavity reference phase $\phi_{\rm rf}$ is usually zero. In the linear approximation, a chicane acts like a drift of length M_{56}

$$\begin{pmatrix} z_2 \\ \delta_2 \end{pmatrix} = \begin{pmatrix} 1 & M_{56} \\ 0 & 1 \end{pmatrix} \begin{pmatrix} z_1 \\ \delta_1 \end{pmatrix} \tag{14.18}$$

Combining these two matrices yields

$$\begin{pmatrix} z_2 \\ \delta_2 \end{pmatrix} = \begin{pmatrix} 1 + M_{65}M_{56} & M_{56}M_{66} \\ M_{65} & M_{66} \end{pmatrix} \begin{pmatrix} z_0 \\ \delta_0 \end{pmatrix} \tag{14.19}$$

The higher-order terms in a more general Taylor expansion for a chicane

$$z_2 - z_1 = M_{56}\delta_1 + T_{566}\delta_1^2 + U_{5666}\delta_1^3 + \cdots \tag{14.20}$$

can also be significant. For a four-dipole chicane

$$\begin{aligned} T_{566} &= -(3/2)\,M_{56} \\ U_{5666} &= \quad 2\,M_{56} \end{aligned} \tag{14.21}$$

so particles with large δ may create nonlinear distortions of the phase space during compression. (See Exercise 14.1).

14.4 Recirculating and Energy Recovery Linacs

The capital cost per unit length of RF structures and RF power sources (and cryomodules and cryogenic systems in the case of superconducting RF) is much higher than the cost of transport lines. Recirculating linear accelerators (RLAs) take advantage of this by adding return arcs to pass the electron beam through the same linac more than once. The effective maximum energy of the linac is multiplied by about the number of turns, at the expense of larger RF power sources. This concept only works effectively for $\beta = 1$ constant frequency structures; GeV-energy protons are recirculated, instead, through swept-frequency RF cavities in synchrotrons.

CEBAF [30] is an example of a 12 GeV RLA in which electron beams pass five times through two 1.09 GeV superconducting linacs that are joined by independent return arcs, as illustrated in Figure 14.3. Multiple beam energies are accelerated simultaneously in each linac. All beams accelerate on-crest and have the same energy gain, but also pass through the same linac quadrupoles. The linac optics, usually FODO, are designed for the lowest energy beam. At the highest energy the quadrupole focusing almost disappears, placing a practical limit on the overall length (and energy gain) of the linac, and on the maximum number of passes.

Many modern applications require superior quality electron beams, for example, for the generation of highly coherent, high average brightness photon beams, for electron cooling devices and for electron colliders for nuclear and particle physics research. As electron accelerators move into the multi-MW ultra-low emittance domain, even RLAs become economically inefficient, at the same time that electron synchrotrons (although energy efficient) are limited in beam quality by the synchrotron radiation and quantum effects discussed in Chapter 12. Energy recovery linacs (ERLs) provide a path forward, towards energy efficient high power operation with high quality beam parameters [32, 53].

Energy Recovery Linacs

The generic layout of a single-pass ERL is shown in Figure 14.4. Beam is injected into a linac and accelerated on-crest as normal, before being recirculated through a

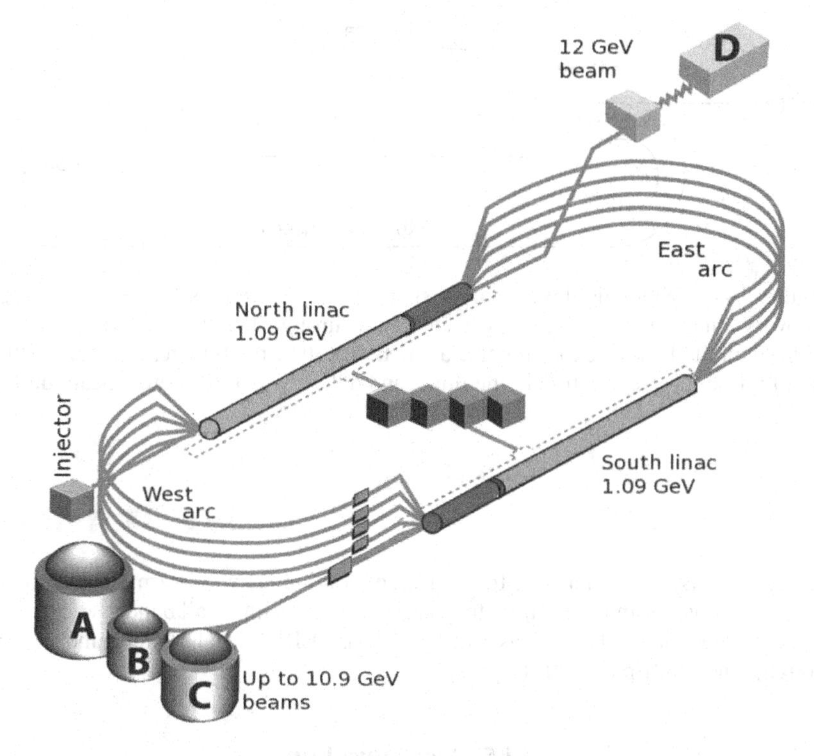

Figure 14.3 The 12 GeV CEBAF recirculating linear accelerator. Beam from the 123 MeV injector is accelerated by 1.09 GeV in the north linac before entering an energy separator that sends five different beam energies to separate return arc transport lines. These beams are recombined before entering the south linac to be accelerated further. Beam is extracted to the experimental halls at any of the five energies.

return arc, as in an RLA. However, the path length is adjusted so that the returning beam is 180° out of phase with the RF. The beam decelerates on its second pass, returning to the RF cavities the energy that it gained in the acceleration pass, finally going to a low power beam dump at an energy comparable to the injection energy.

It is natural to use superconducting cavities in ERLs, since they help in the attainment of high drive power efficiency. Ideally the required RF drive power barely changes with beam power, permitting economical production of multi-MW beams for experimental uses, so long as the experiments do not interfere significantly with energy recovery by unduly disrupting the recirculating beam intensity and quality.

The ERL concept can also be extended to multiple passes. Bunch patterns for various single-pass and multi-pass linac configurations are sketched in Figure 14.5.

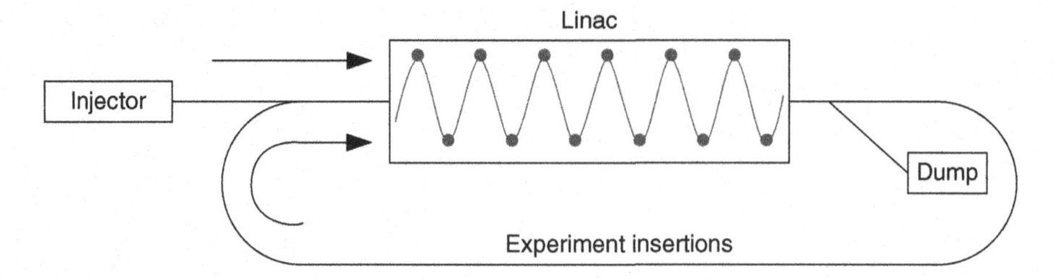

Figure 14.4 Schematic layout of a single-pass continuous waveform energy recovery linac. Injected beam is accelerated through the linac on-crest, before being exploited in an experimental area. Beam then recirculates to return 180° out of phase for deceleration in the linac, and delivery to a low power beam dump.

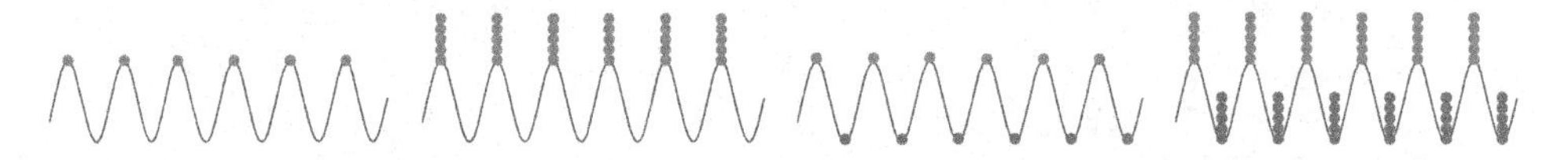

Figure 14.5 Bunch patterns in four different continuous waveform electron linac configurations. From left to right, they are: single-pass linac (all bunches on-crest), recirculating linac (RLA, 5 passes), single-pass ERL (decelerating bunches on-trough), and multi-pass ERL (5 passes).

14.5 Beam Breakup

Beam breakup (BBU) instabilities are the most common instabilities of concern in electron linacs [5, 57]. They occur due to interactions of off-axis electrons with the higher-order modes of RF cavities. Excited modes driven by the beam persist for several damping times, interacting with electrons that pass later – in the same or different bunches, on the same or subsequent passes.

Consider the two-macroparticle head-tail model of a single bunch. The head macroparticle of charge $Ne/2$ executes transverse betatron oscillations with an amplitude $\hat{x}$

$$x_{\text{head}}(s) = \hat{x}\cos(k_\beta s) \tag{14.22}$$

as s increases along the linac, where k_β is the wave number of the oscillation. The tail macroparticle, a distance z behind the head, performs the same betatron oscillation with the addition of a driving term due to its interaction with higher-order modes driven by the head macroparticle. In a smooth approximation

$$x''_{\text{tail}}(s) + k_\beta^2\, x_{\text{tail}}(s) = -\left(\frac{Ne^2 W_1(z)}{2EL}\right)\hat{x}\cos(k_\beta s) \tag{14.23}$$

where E is the electron energy, L is the period between RF cavities and $W_1(z)$ is the transverse wake function that depends on the higher-order modes and on the time

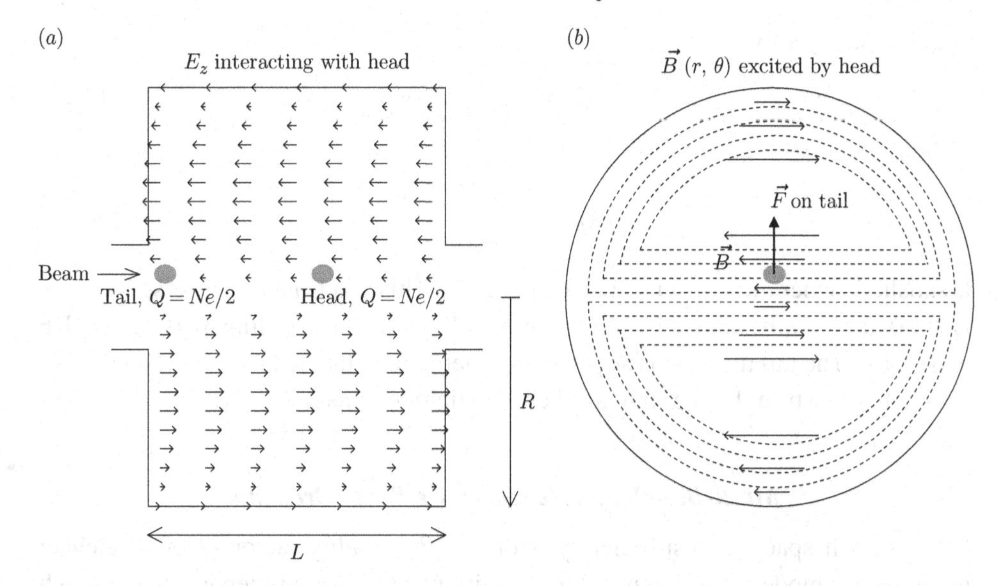

Figure 14.6 Beam breakup interaction of an electron bunch with an RF cavity TM_{110} higher-order mode. The transverse displacement of the head and tail macroparticles is shown in (a), the head of the bunch decelerates and deposits energy into the HOM. The displaced tail of the bunch is shown in (b), interacting with the HOM's magnetic field to produce a transverse force $\vec{F}$. This feedback mechanism may be anti-damped, driving an instability. Compare to Figure 7.4.

(or distance z) since the modes were excited. The beam breakup mechanism for a TM_{110} HOM is sketched in Figure 14.6. The solution of Equation 14.23

$$x_{\text{tail}}(s) = \hat{x}\cos(k_\beta s) - \left(\frac{Ne^2 W_1(z)}{4k_\beta EL}\right) s\,\hat{x}\,\sin(k_\beta s) \tag{14.24}$$

has a tail amplitude that grows linearly with s. This growth increases the effective transverse emittance of the electron bunch, with the potential to cause beam loss in long linacs.

Acceleration along the linac has so far been neglected. When it is included, the tail amplitude growth is found to be logarithmic in s, rather than linear [5]. Even this growth is quite severe for long linacs. The constraint on the initial trajectory amplitude $\hat{x}$ can still be intolerable, without further compensation.

Single-bunch beam breakup is controlled by the BNS damping mechanism (due to Balakin, Novokhatsky and Smirnov), which introduces an additional betatron focusing δk_β that acts on only the tail of the bunch [1]. Equation 14.23 then becomes

$$x''_{\text{tail}}(s) + (k_\beta + \delta k_\beta)^2\, x_{\text{tail}}(s) = -\left(\frac{Ne^2 W_1(z)}{2EL}\right)\hat{x}\cos(k_\beta s) \tag{14.25}$$

which is safely solved by

$$x_{\text{tail}}(s) = \hat{x}\cos(k_\beta s) \tag{14.26}$$

if the BNS condition

$$\delta k_\beta = -\frac{Ne^2 W_1(z)}{4\, k_\beta\, EL} \tag{14.27}$$

is satisfied, under the assumption that $\delta k_\beta \ll k_\beta$. This additional transverse focusing is most commonly added by accelerating off-crest, on a falling part of the RF waveform. The tail then gains slightly less energy than the head, and is transversely focused more strongly due to natural chromaticity (Chapter 9).

Multi-Bunch and Regenerative Beam Breakup

When bunch spacing is sufficiently short, or the quality-factor Q of a relevant higher-order mode is sufficiently high, cavity excitations can persist long enough to interact with successive bunches in the linac. Multi-bunch beam breakup is a particular concern in high repetition rate electron linacs, and in high power proton linacs with superconducting RF.

Multi-bunch beam breakup is much more complicated than the single-bunch example given above. Each RF cavity has many HOMs, both longitudinal and transverse. Successive bunches interact with HOMs at different phases and times in their decay, exchanging power with the HOMs, and enabling different HOMs to couple and exchange power. Computer modelling is necessary to predict beam and HOM behaviour in large superconducting linacs, and to determine HOM damping constraints when designing the RF systems.

Regenerative beam breakup is a threat when an electron beam recirculates through the same linac more than once, in an RLA or an ERL. A bunch deflected by an HOM on its first pass may be displaced off-axis on its second pass, further exciting the HOM and creating a closed feedback loop between the HOM and the beam. This feedback becomes unstable if the beam current exceeds an instability threshold that depends on the relative timing or phase of the recirculated beam relative to the HOM. In some cases recirculating beam damps the HOM excitation, rather than driving it.

The threshold current for horizontal regenerative beam breakup is

$$I_{\text{th}} = -\frac{2Ec}{e\omega\,(R/Q)Q\,M_{12}\,\sin(\omega T)} \tag{14.28}$$

where E is the total beam energy, ω is the HOM frequency, Q and (R/Q) are properties of the HOM, M_{12} is the recirculation beam transport matrix element, and T is the recirculation time [45]. Instability is only possible if

$$M_{12} \sin(\omega T) < 0 \tag{14.29}$$

Otherwise the feedback loop damps the HOM, and the beam is predicted to remain stable at all currents. Regenerative beam breakup can be avoided for some HOMs by careful control of M_{12}. However, $\omega T \gg 1$ in realistic systems so stability cannot be assured for all HOMs.

Exercises

14.1 Figure 14.7 sketches the layout of a simple four-dipole chicane in which the bend angle θ is small, and a is the distance between the first and second dipoles.

Figure 14.7 A simple four-dipole chicane, with no quadrupoles.

a) Following Equation 14.18, show that

$$M_{56} \equiv \frac{dz_2}{d\delta_1} = -2a\theta^2 \tag{14.30}$$

b) Following the Taylor expansion of the path length in Equation 14.20, show that

$$T_{566} = -\frac{3}{2}M_{56} \tag{14.31}$$

and

$$U_{5666} = 2M_{56} \tag{14.32}$$

c) At the front of the bunch is z negative or positive?

14.2 Derive the BNS condition (Equation 14.27) by solving the equation of motion (Equation 14.25).

15

The Beam–Beam Interaction and 1-D Resonances

The beam–beam interaction has long been the subject of intense study, not only because it fundamentally limits the luminosity of electron–positron and hadron–hadron colliders, but also because it exhibits complex dynamical behaviour, even in its simplest abstraction. Consistent with Taff's comment about 'beyond first-order [knowing] ... no useful result from perturbation theory' [52], first-order theory does an excellent job of describing beam–beam behaviour. The second-order theory that is often necessary in describing magnetic nonlinear behaviour (for example in the presence of chromaticity-correcting sextupoles) is at best complex, and sometimes nonsensical. The simple case of a single round beam–beam interaction, presented here, is an instructive and accessible example of 1-D resonant behaviour that is analytically soluble in closed form even at large amplitude, and which is quantitatively accurate.

We study magnet nonlinearities (in part) for economic reasons: how to minimise costs while guaranteeing an adequately large transverse aperture for stable betatron oscillations? In contrast we study the beam–beam interaction to probe fundamental limits: why does the beam–beam parameter ξ (yet to be defined) have maximum values of

$$\begin{aligned} e^+ - e^- \quad \text{storage ring}: \quad \xi_{max} &\sim 0.1 \\ h - h \quad \text{storage ring}: \quad \xi_{max} &\sim 0.01 \end{aligned} \tag{15.1}$$

in the two kinds of collider? How can collider designs be tuned to increase this limit by even just a few per cent, with a corresponding increase in single bunch intensity and luminosity?

15.1 Round Beam-Beam Interaction

Figure 15.1 illustrates a negatively charged test particle passing through the magnetic and electric fields of a cylindrically symmetric positively charged bunch, with a charge distribution of

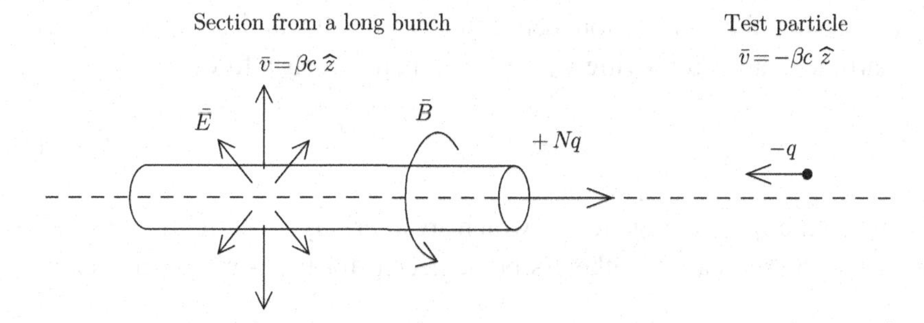

Figure 15.1 A negatively charged test particle entering a beam–beam 'collision' is about to pass through the electrostatic and magnetic fields of a positively charged counter-rotating bunch that is long, thin and round.

$$\rho = \rho(z) \cdot \frac{e^{-r^2/2\sigma^2}}{2\pi\sigma^2} \tag{15.2}$$

where the horizontal and vertical RMS beam sizes are equal $\sigma = \sigma_x = \sigma_z$ and the total charge is

$$Nq = \int_{-\infty}^{\infty} \rho(z)\, dz \tag{15.3}$$

The force on the test particle is

$$\bar{F} = -q\left(\bar{E} + \bar{v} \times \bar{B}\right) \tag{15.4}$$

where both force components are inwardly focusing.

If the bunch is much longer than its width $\sigma_z \gg \sigma$ then the forces are purely transverse and the electric force is

$$F_E = -qE = \frac{q^2}{2\pi\epsilon_0}\frac{n(r)}{r} \tag{15.5}$$

where, using Gauss' law,

$$n(r) = \rho(z)\left[1 - e^{-r^2/2\sigma^2}\right] \tag{15.6}$$

Similarly it can be shown that the magnetic force is

$$F_B = \beta^2 F_E \tag{15.7}$$

so that the total transverse force is

$$F_\perp = (1 + \beta^2) F_E \tag{15.8}$$

Note in passing the connection between beam–beam and space charge forces. A test particle that is co-moving with its own bunch feels a force

$$F_{\perp,SC} = (1-\beta^2)F_E = \frac{F_E}{\gamma^2} \tag{15.9}$$

showing that the space charge forces which are a serious concern for non-relativistic beams when $\beta \approx 0$ can nonetheless be neglected for ultra-relativistic beams with $\gamma \gg 1$.

The beam–beam force of Equation 15.8 is readily integrated along the collision length if the β-function is much larger than the bunch length $\beta^* \gg \sigma_z$, so that the transverse beam size and betatron phase are approximately constant throughout the interaction. In that case the net horizontal angular kick received by a particle that is horizontally displaced by x is

$$\begin{aligned}\Delta x' &= \int \frac{F_\perp}{p_0}\,dt = \frac{1}{2p_0c}\int_{-\infty}^{\infty} F_\perp\,dz \\ &= \frac{-Nq^2}{2\pi\epsilon_0\,mc^2(\beta\gamma)}\cdot\frac{1}{x}\left(1-e^{-x^2/2\sigma^2}\right)\end{aligned} \tag{15.10}$$

which is simplified by introducing the beam–beam parameter ξ so that

$$\Delta x' = -\frac{4\pi\xi}{\beta^*}\cdot\frac{2\sigma^2}{x}\left(1-e^{-x^2/2\sigma^2}\right) \tag{15.11}$$

as illustrated in Figure 15.2. At large displacements $x \gg \sigma$ the beam–beam kick converges to zero like $1/x$ for any beam distribution, in stark contrast to magnetic kicks which diverge to infinity like x^m, as shown for $m = 3$ (an octupole) in Figure 15.2. This convergence allows first-order theory to be accurate even for large amplitude particles.

Since the size of a beam with an RMS normalised emittance of ϵ_N is

$$\sigma = \sqrt{\frac{\epsilon_N\beta^*}{(\beta\gamma)}} \tag{15.12}$$

and recalling that the classical radius of a proton (or electron) is

$$r_0 = \frac{q^2}{4\pi\,\epsilon_0\,mc^2} \tag{15.13}$$

then the value of the dimensionless beam–beam parameter is simply

$$\xi = \frac{Nr_0}{4\pi\,\epsilon_N} \tag{15.14}$$

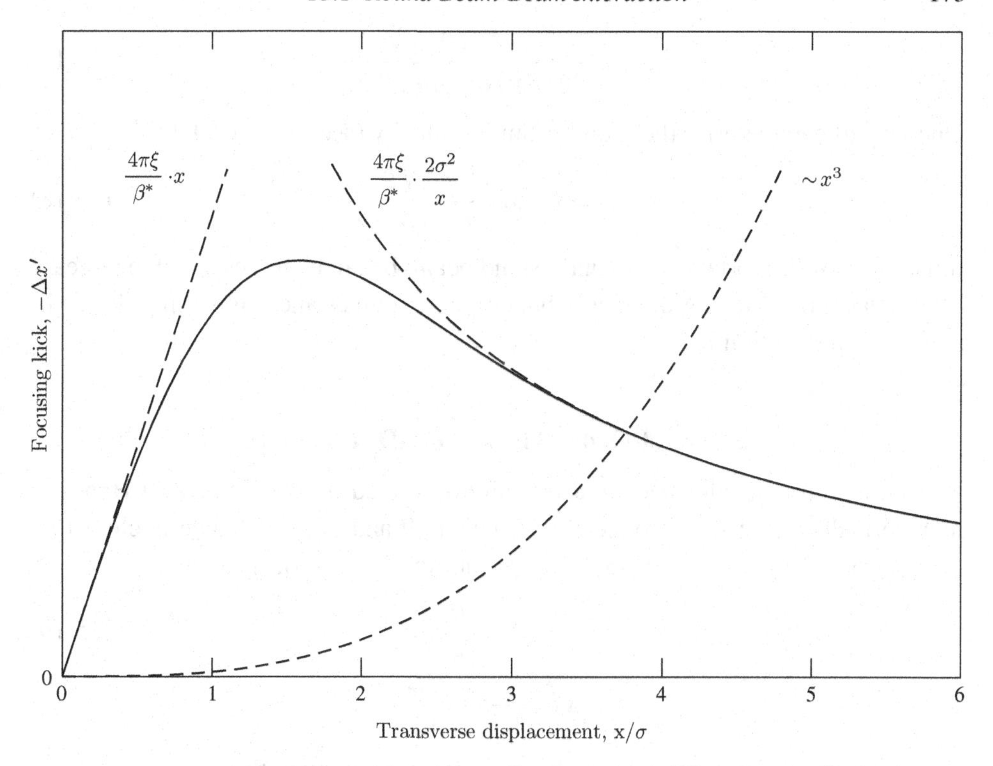

Figure 15.2 The focusing kick of a round beam–beam interaction described by Equation 15.11. The kick converges to zero like $1/x$ at large displacements, unlike magnetic kicks which diverge like x^m.

for a round beam–beam collision. At small displacements $x \ll \sigma$ the interaction behaves like a focusing quadrupole with a focal length f

$$\Delta x' \approx -\frac{4\pi\xi}{\beta^*}\cdot x = -\frac{1}{f}\cdot x \tag{15.15}$$

Applying Equation 8.29 shows that the betatron tune increases by

$$\Delta Q = \frac{\beta^*}{4\pi}\cdot\frac{1}{f} = \xi \tag{15.16}$$

for small amplitude oscillations, explaining why the beam–beam parameter is often called the ‘beam-beam tune-shift parameter’. It is striking that this tune shift is independent of both optics β^* and energy γ (see Equation 15.14).

Life is only a bit more complicated in the general bi-Gaussian case, when the beam is not round and the β-functions are not equal at the collision point. In that case the horizontal and vertical beam–beam parameters, and small amplitude tune shifts, become

$$\xi_{H,V} = \frac{N r_0 \, \beta_{H,V}}{2\pi (\beta\gamma) \, \sigma_{H,V}(\sigma_H + \sigma_V)} \tag{15.17}$$

One way to parameterise the round beam luminosity (see Equation 1.11)

$$L = f_{rev} \, M \, \frac{(\beta\gamma)}{r_0} \, \frac{N \xi}{\beta^*} \tag{15.18}$$

illustrates how important it is to understand beam–beam dynamics – and to increase ξ while also maximising the single bunch population N and minimising the collision point β-function β^*.

15.2 First-Order Theory of 1-D Resonances

Motion in action–angle space is accurately described by the Kobayashi Hamiltonian formalism already introduced in Chapters 9 and 10, if the tune is close to a resonance $Q_0 \approx p/n$ [25, 40, 43]. The net motion over n turns is then

$$\Delta\phi = n \cdot \frac{\partial H_n}{\partial J} \tag{15.19}$$
$$\Delta J = -n \cdot \frac{\partial H_n}{\partial \phi}$$

where in general for a single beam-beam interaction of strength ξ

$$H_n = 2\pi \left(Q_0 - \frac{p}{n}\right) J \; + \; 2\pi\xi \; U(J) \; - \; 2\pi\xi \; V_n(J) \cos(n\phi) \tag{15.20}$$

It is relatively easy to extend this Hamiltonian to include multiple collisions per turn, and a second transverse dimension.

In the particular case of a single round collision the detuning and resonance functions U' and V_n' are

$$U'(J) = \frac{2}{J} \left[1 - e^{-J/2} \, I_0(J/2)\right] \tag{15.21}$$
$$V_n'(J) = (-1)^{n/2} \left(\frac{4}{J}\right) e^{-J/2} \, I_{n/2}(J/2)$$

where I_m is a modified Bessel function of the first kind, and a prime indicates differentiation with respect to J [24, 43]. The betatron amplitude is related to the (conveniently scaled and dimensionless) action through

$$a = \sqrt{2J} \, \sigma \tag{15.22}$$

Note that $V_n'(J)$ is zero when n is odd.

The betatron phase advance from turn t to $t+n$ is

$$\phi_{t+n} \; - \; \phi_t = 2\pi n \left[Q_0 + \xi \, U'(J) \; - \; \xi \, V_n'(J) \cos(n\phi)\right] \tag{15.23}$$

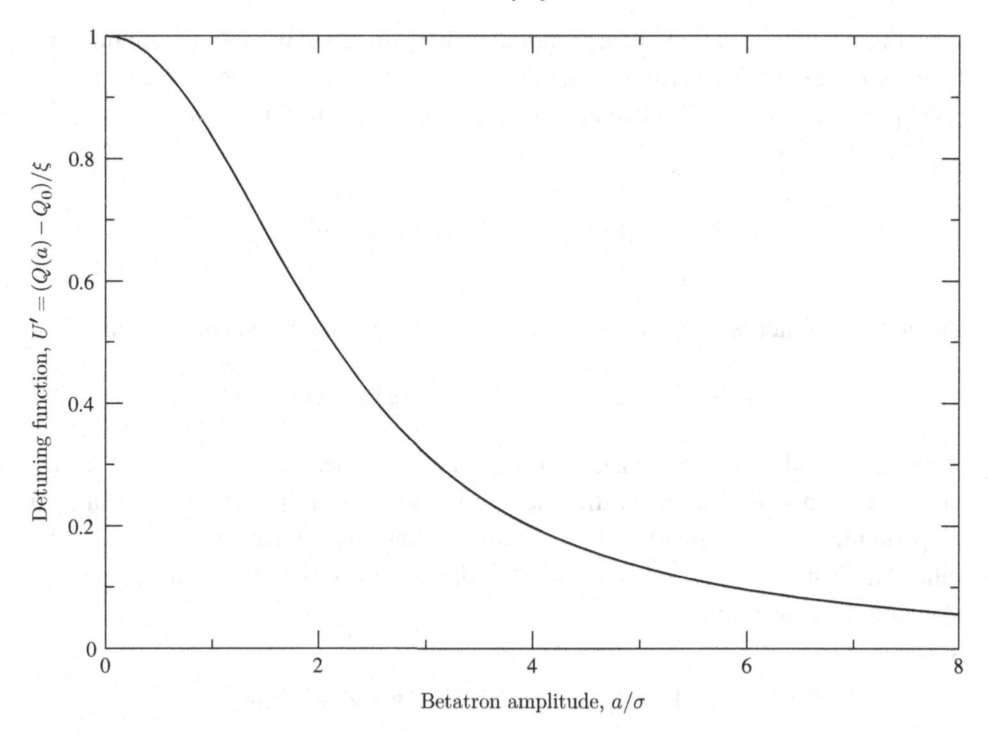

Figure 15.3 The detuning function U' for a single round beam–beam collision per turn.

so that the tune – the average phase advance per turn (divided by 2π) – is

$$Q(J) \equiv \frac{1}{2\pi n} \langle \phi_{t+n} - \phi_t \rangle \tag{15.24}$$

In the approximation that ϕ is evenly distributed in the range 0 to 2π, so that $\cos(n\phi)$ averages to zero, then the tune varies with action simply as

$$Q(J) = Q_0 + \xi\, U'(J) \tag{15.25}$$

which explains why U' is called the detuning function. Figure 15.3 shows how U' and the tune shift vary with the betatron amplitude a, confirming that the tune shift is ξ in the small amplitude limit.

The n-turn motion described by Equations 15.19 and 15.20 includes fixed points: phase space locations with action–angle co-ordinates that exactly repeat themselves after n turns. Fixed points are located at local minima, local maxima or saddle points of H_n. Stable and unstable motion near them is described by introducing the resonant action J_R that depends on Q_0 and ξ through

$$Q = Q_0 + \xi\, U'(J_R) = \frac{p}{n} \tag{15.26}$$

According to this equation, the *nominal* tune at J_R (in the now-false assumption that ϕ is evenly distributed between 0 and 2π) is exactly on-resonance.

Expanding around J_R with small values of I gives approximations

$$J \equiv J_R + I \tag{15.27}$$
$$U(J) \approx U(J_R) + U'(J_R)I + \frac{1}{2}U''(J_R)I^2$$
$$V_n(J) \approx V_n(J_R)$$

that are substituted into Equation 15.20 to derive the resonance Hamiltonian

$$H_{Rn}(\phi, I) = 2\pi \left[\frac{1}{2} (\xi U_R'') I^2 - (\xi V_{Rn}) \cos(n\phi) \right] \tag{15.28}$$

where $\xi U_R''$ and ξV_{Rn} are constants for particular values of J_R (and hence of Q_0 and ξ). The only fundamental difference between this Hamiltonian and the standard map Hamiltonian of Equation 4.40 is that this has an angular term $n\phi$, while the other has just q. In general, there are n islands in a resonance chain, while the standard map has only one.

15.3 Resonance Island Tunes and Widths

Difference motion near a resonance is described by differentiating the resonance Hamiltonian

$$\phi_{t+n} - \phi_t = n \frac{\partial H_{Rn}}{\partial I} \tag{15.29}$$
$$I_{t+n} - I_t = -n \frac{\partial H_{Rn}}{\partial \phi}$$

In a differential (continuous time) approximation this motion becomes

$$\frac{d\phi}{dt} = 2\pi \, \xi U_R'' \cdot I \tag{15.30}$$
$$\frac{dI}{dt} = -2\pi \, n \xi V_{Rn} \cdot \sin(n\phi)$$

These equations are fundamentally the same as the standard map introduced in Section 4.4, when $n = 1$. In general there are n stable and n unstable fixed points, at

$$I_{FP} = 0 \tag{15.31}$$
$$\cos(n\phi_{FP}) = \pm 1$$

A six-island chain due to a single round beam–beam collision is visible in Figure 15.4, for two different values of ξ with

$$Q_0 < \frac{2}{6} < Q_o + \xi \tag{15.32}$$

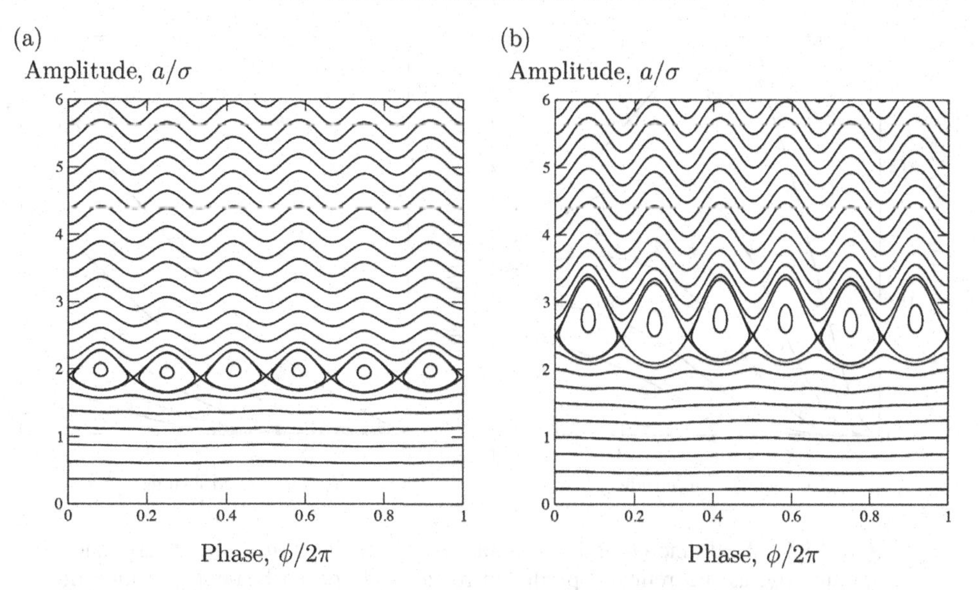

Figure 15.4 Six-island chains from the simulation of a single round beam-beam interaction of strength $\xi = 0.0042$ (a), and $\xi = 0.006$ (b), with a base tune of $Q_0 = 0.331$ [40]. The amplitude width of the islands increases as the chain moves to a larger resonance amplitude when ξ is increased. (See also Figure 16.2.)

and with island centres at resonance amplitudes of about 2.0 σ and 2.7 σ. The trajectories of dense turn-by-turn dots appear (mainly) as solid lines, following contours of the resonance Hamiltonian $H_{R6}(\phi, I)$.

Stable motion trajectories circulate the fixed points at resonance island centres with $\cos(n\phi_{FP}) = -1$, according to Equation 15.30. Small linear oscillations close to an island centre behave like

$$I = a_I \cos(2\pi Q_I t) \tag{15.33}$$
$$\phi = \phi_{FP} + a_\phi \sin(2\pi Q_I t)$$

where the island tune derived from Equation 15.30

$$Q_I = n\xi\,(-V_{Rn} U_R'')^{1/2} \tag{15.34}$$

is directly proportional to the beam–beam parameter ξ. The ratio Q_I/ξ is in the range from 0 to 0.6 for beam–beam resonances from order $n = 4$ to 12, as shown quantitatively in Figure 15.5. Small stable oscillations only access phases in the range

$$\phi_{FP} - a_\phi \;<\; \phi \;<\; \phi_{FP} + a_\phi \tag{15.35}$$

so the motion is phase-locked, and the tune is exactly p/n!

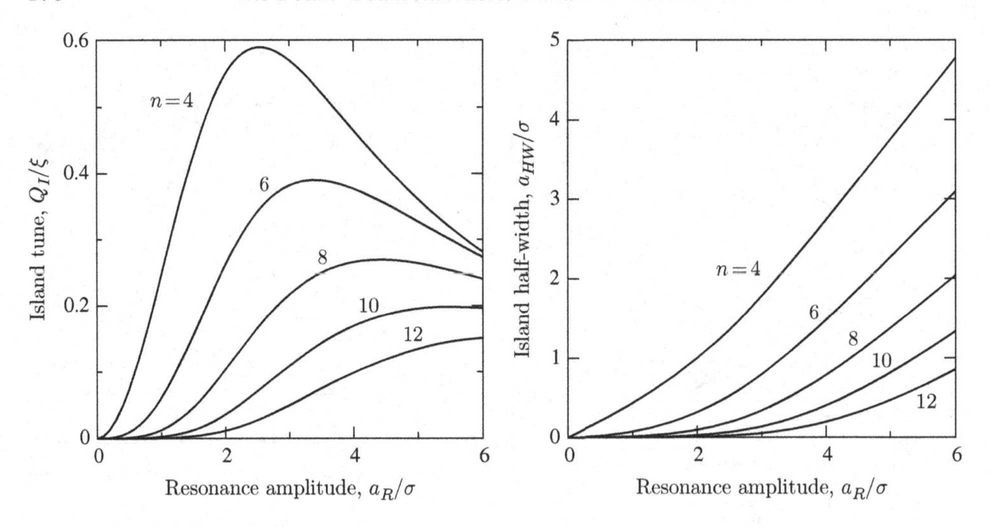

Figure 15.5 Island tune Q_I and island half-width a_{HW} as a function of resonance amplitude a_R, as theoretically predicted for a single round beam–beam kick of strength ξ, for resonances from order $n = 4$ to 12.

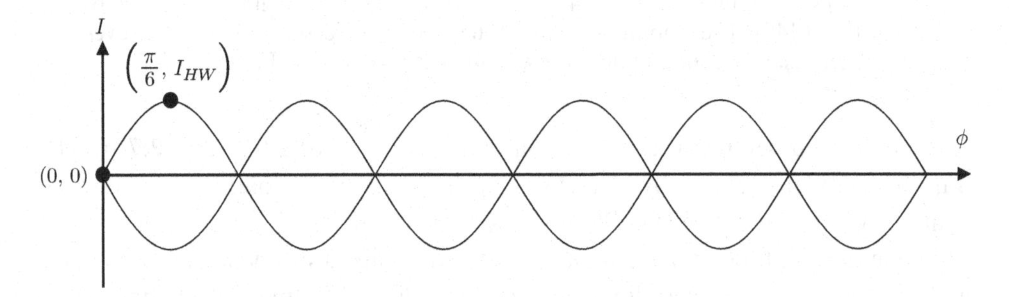

Figure 15.6 The idealised separatrix contours of a six-island chain. Comparing the separatrix values of the Hamiltonian H_{Rn} at stable and unstable fixed point phases enables a calculation of the island half-width I_{HW}.

Unstable motion launched close to a fixed point with $\cos(n\phi_{FP}) = +1$ reaches all values of phase ϕ, travelling just outside the separatrix contours that skirt the entire island chain, as sketched in Figure 15.6. The island half-width I_{HW} is calculated by equating the values of the Hamiltonian contour on the separatrix at the phases of stable and unstable fixed points

$$H_{Rn}(0,0) = H_{Rn}\left(\frac{\pi}{6}, I_{HW}\right) \tag{15.36}$$

Invoking Equation 15.28 leads to the remarkable result

$$I_{HW} \approx 2\left(\frac{V_{Rn}}{U_R''}\right)^{1/2} \tag{15.37}$$

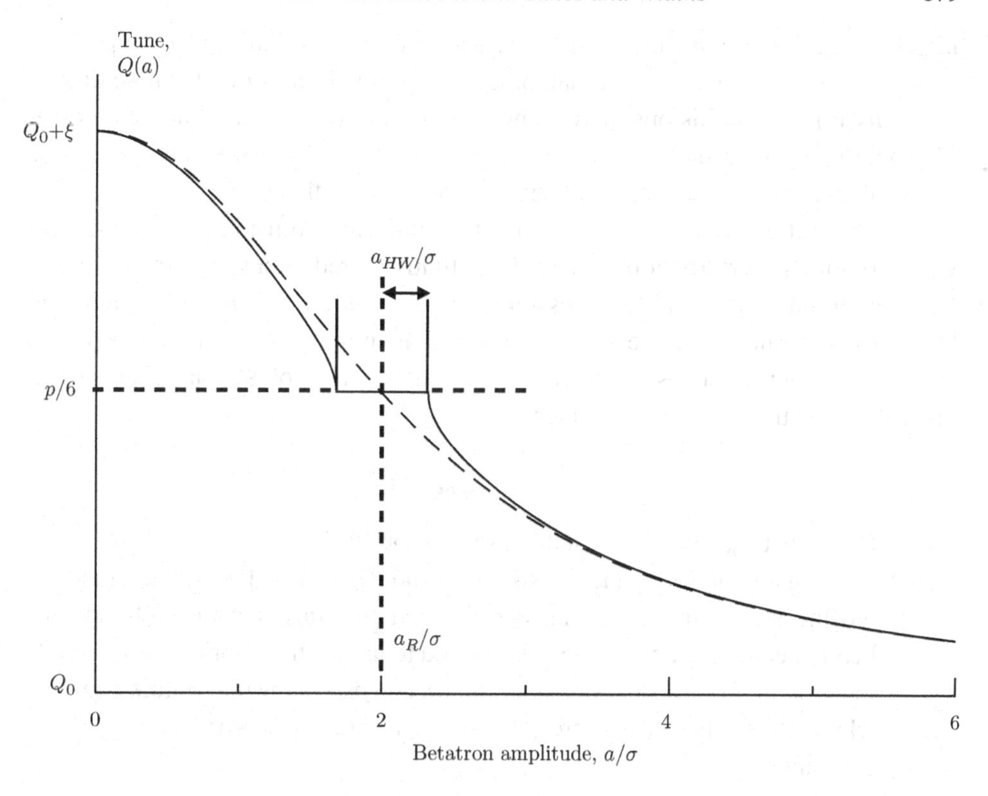

Figure 15.7 Tune versus amplitude for a set of particles launched with $\phi = \phi_{FP}$, when Q_0 and ξ values cause the shifted tune to cross a sixth-order resonance nominally at $a_R/\sigma = 2$. Particles with amplitudes in the range $(2.0 \pm 0.317)\,\sigma$ are phase locked, and have a tune of exactly $p/6$.

that the island width is independent of the beam–beam parameter! For example, Figure 15.7 shows that a sixth-order resonance has an island half-width of $a_{HW} = 0.317\,\sigma$ when $a_R = 2.0\,\sigma$, independent of ξ.

One way to understand this independence is by considering that the detuning strength $\xi U''$ and the resonance driving strength ξV compete to decrease and increase the width. Since both are proportional to the beam–beam parameter, the net dependence on ξ disappears. Figure 15.5 shows that island half-widths grow monotonically with resonance amplitude a_R, and are comparable to σ even for twelfth-order resonances.

Flat Beams and Synchrotron Radiation

In practice, colliders usually have more than one collision per turn, the beams are not perfectly round and there are two more dimensions to consider. Electron–positron colliders have flat beams, with the emittance, β^* and beam size all much larger in the horizontal than in the vertical. More important, electron–positron collisions benefit from radiation damping, which ameliorates the dynamic disruption caused by the collisions, permitting beam–beam parameters ξ about an order of magnitude larger than in hadron–hadron colliders. Unfortunately these vital practical and nuanced concerns are beyond the scope of this book.

The abstract example of a single round beam–beam collision is fully tractable with first-order perturbation theory. Quantitative predictions out to arbitrarily large amplitudes are readily accessible for verification by simple simulations. Single round beam–beam theory also enables a quantitative introduction to chaotic behaviour. When is chaos expected? With which values of ξ? Via what mechanisms? This is the domain of Chapter 16.

Exercises

15.1 Prove that $F_B = \beta^2 F_E$ as stated in Equation 15.7.

15.2 Investigate motion under a single round Gaussian 1-D interaction by writing simulation code that permits you to adjust the tune Q and the beam–beam parameter ξ, and allows you to launch trajectories at any initial location in phase space. Consider the (a, ϕ) space shown in Figure 15.4, in which almost-flat lines correspond to regular resonance-free motion, with detuning.

(a) Set $Q = 0.331$ and $\xi = 0.006$, and observe the resonance islands that appear at an amplitude where the beam-beam tune shift moves particles across the $Q = 2/6$ resonance. Compare the amplitude of the resonance island centres to the theoretical prediction.

(b) Why does only every second resonance island appear?

(c) Motion near $Q = 1/3$ becomes unbounded at modest amplitudes when significant sextupoles are present, but here the motion is regular for even the largest amplitudes. Why?

(d) Set the tune to $Q = 0.305$ and print (or save) phase space diagrams for four or five values of ξ in the range from 0.05 to 0.2. What happens as ξ increases?

15.3 Use a numerical simulation to test the dependence of island tune Q_I on resonance amplitude a_R/σ that is predicted in Equation 15.34 and plotted in Figure 15.5. For example, set Q to a series of values near 1/3 (or 1/4 or 2/5), then measure a_R and the period of small oscillations around the centre of one of the resonance islands.

15.4 The luminosity in electron–positron colliders varies as I^2 at low currents. At high currents the luminosity is often linear in I. Account for this beam–beam behaviour.

16

Routes to Chaos

The standard map emerges when the differential equation of a pendulum is translated to a pair of difference equations, as shown in the discussion of longitudinal motion and synchrotron oscillations in Section 4.4. Its only control parameter, Q_0, is analogous to the synchrotron tune of small oscillations in a storage ring with a single RF cavity – or with a set of cavities that are close enough to act as one. The rich phase space behaviour that emerges as Q_0 increases is shown in Figure 4.6. A small chaotic area emerges when $Q_0 = 0.12$, easily identifiable as a scattering of dots. Stable resonance islands emerge from a broad chaotic sea in the extreme case with $Q_0 = 0.18$. Stability and chaos are intermingled at all scales – stability within chaos, chaos within stability – ad infinitum. This is what Hofstadter means, when he refers to '*. . . an eerie type of chaos . . . just behind a facade of order - and yet, deep inside the chaos lurks an even eerier type of order*' [20].

The Hénon map is analogous to a storage ring with a single sextupole, as discussed in Section 9.3. This map generates the rich horizontal behaviour shown graphically in Figure 9.3. The taxonomy of 1-D motion discussed in Section 9.4 categorises turn-by-turn phase space trajectories as:

1. Regular non-resonant;
2. Rapidly divergent;
3. Regular resonant; or
4. Chaotic.

This taxonomy is universal. It emerges in phase space motion driven by one sextupole or many, by beam–beam interactions, or by RF cavities.

Much of this book addresses practical and quantitative aspects of linear and near-linear *regular non-resonant* (non-chaotic) motion. *Rapidly divergent* trajectories are discussed semi-quantitatively in Chapter 10, in the practical context of slow extraction from a storage ring. *Regular resonant* motion is discussed and quantified in Chapter 15, using the convenient and fundamental example of the 1-D beam–beam interaction.

Chaos can also be predicted and discussed semi-quantitatively, beyond the qualitative observation that it appears as dots in phase space, and beyond its formal definition using Lyapunov exponents. The routes to chaos examined in this chapter inevitably require more than just one real-space dimension, especially when more practical and realistic situations are addressed.

16.1 Resonance Overlap

The theory of 1-D resonances introduced in Section 15.2 uses the round beam–beam interaction as a convenient example, summarised by the n-turn Kobayashi Hamiltonian

$$H_n = 2\pi\left(Q_0 - \frac{p}{n}\right) J \;+\; 2\pi\xi\, U(J) \;-\; 2\pi\xi\, V_n(J)\cos(n\phi) \tag{16.1}$$

where n is the order of the resonance, ξ is the beam–beam parameter, $U' = dU/dJ$ is the detuning function, V_n is the resonance driving function, and the action J is scaled to the betatron amplitude a through

$$a = \sqrt{2J}\,\sigma \tag{16.2}$$

where σ is the RMS size of the beam. In contrast, the n-turn Hamiltonian for near-resonance motion dominated by many nonlinear magnets is well approximated by a simple polynomial. For example, close to a fifth order resonance driven only by sextupoles

$$H_5 = 2\pi\left(Q_0 - \frac{p}{5}\right) J \;+\; \alpha J^2 \;+\; V_5\, J^{5/2}\cos(5\phi + \phi_5) \tag{16.3}$$

where V_5 and ϕ_5 are constants [40]. Comparison to Equation 10.18 shows that the constant α measures the strength of the octupolar tune variation with amplitude since, away from the resonance,

$$Q = Q_0 + \left(\frac{\alpha}{\pi}\right) J = Q_0 + \left(\frac{\alpha}{2\pi\sigma^2}\right) a^2 \tag{16.4}$$

in analogy to Equation 10.9. Even sextupoles generate octupolar detuning!

No matter whether a 1-D resonance is driven by beam–beam interactions or by nonlinear magnets – or both – the key quantities of interest are the island tune and the island half-width, which vary with the resonance action J_R according to

$$\begin{aligned} Q_I &= n\xi\,\left(-V_n(J_R)U''(J_R)\right)^{1/2} && \text{beam–beam} \\ &= \frac{5}{2\pi}\,(2\alpha V_5)^{1/2}\, J_R^{5/4} && \text{sextupoles } (n = 5) \end{aligned} \tag{16.5}$$

and

$$I_{HW} = 2\left(\frac{V_n(J_R)}{U''(J_R)}\right)^{1/2} \qquad \text{beam–beam} \tag{16.6}$$

$$= \left(\frac{2V_5}{\alpha}\right)^{1/2} J_R^{5/4} \qquad \text{sextupoles } (n = 5)$$

In the beam–beam case the island tune Q_I has a maximum at intermediate values of the resonance amplitude a_R, while the island width I_{HW} is (remarkably) independent of the beam–beam parameter ξ, as shown in Figure 15.5. In the sextupole case both Q_I and I_{HW} scale like $a_R^{5/2}$, while the independent detuning and resonance driving coefficients, α and V_5, compete in determining the island tune half-width.

It has long been understood that if the dynamical control parameters – like (Q_0, ξ) or (Q_0, α, V_5) – are such that two resonance island chains overlap, then massive chaos results [9]. However, the simplest model of resonance overlap predicts beam–beam parameter limits that are unrealistically large. For example, the islands associated with the moderately high order beam–beam resonances 6/10 and

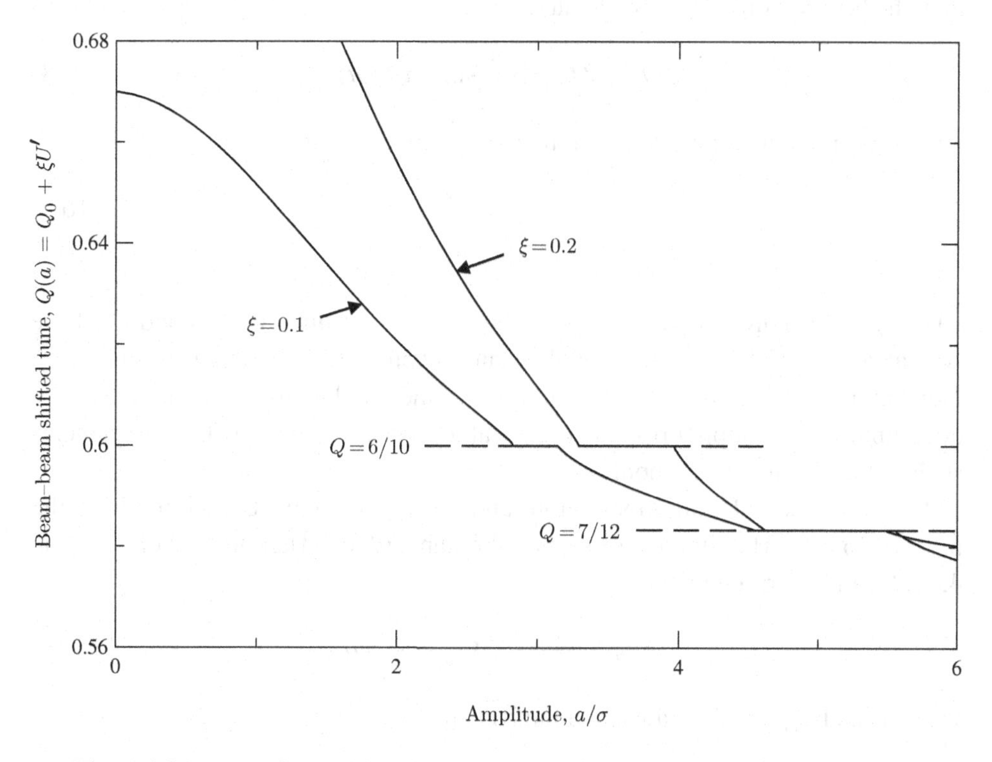

Figure 16.1 Beam–beam driven islands on the 6/10 and 7/12 resonances approach each other, almost overlapping, as the detuning curve steepens when the beam–beam parameter ξ is increased from 0.1 to 0.2. The island widths are independent of ξ, but increase with the resonance amplitude.

7/12 only begin to overlap at around $\xi = 0.2$, as illustrated by Figure 16.1. This is about an order of magnitude larger than the practical limits observed in hadron colliders. Clearly, the simplest 1-D model of beam–beam interactions is deficient.

In reality all six phase space dimensions are coupled together, both into and out of each of the three real space dimensions. For example, off-momentum particles are displaced horizontally in proportion to the dispersion function, and rolled quadrupoles couple horizontal and vertical betatron oscillations.

16.2 Tune Modulation

Betatron tune modulation is driven by synchrotron oscillations, providing a powerful coupling mechanism that is a route to chaos in 1.5 dimensions. If the momentum oscillates with an amplitude $\widehat{\delta}$

$$\frac{\Delta p}{p} = \widehat{\delta} \sin(2\pi Q_S t) \tag{16.7}$$

then the betatron tune Q_0 is modulated

$$Q_0 = Q_{00} + q \sin(2\pi Q_M t) \tag{16.8}$$

with a modulation amplitude and a modulation tune that are given by

$$\begin{aligned} q &= \chi \widehat{\delta} \\ Q_M &= Q_S \end{aligned} \tag{16.9}$$

where χ is the chromaticity and Q_S is the synchrotron tune. Both q and Q_M have typical values of a few parts per thousand when tune modulation is driven by momentum oscillations. Other sources of tune modulation include dipole or quadrupole power supply ripple, perhaps at a spectrum of modulation frequencies such as line frequency harmonics.

In the absence of tune modulation, and by analogy with Equation 15.28, the n-turn resonance Hamiltonian resembles the standard map Hamiltonian of Equation 4.40. It can be written in general as

$$H_{Rn} = \alpha I^2 - V_{Rn} \cos(n\phi) \tag{16.10}$$

where α and V_{Rn} are constants, and the island action

$$I = J - J_R \tag{16.11}$$

is the deviation from the resonance action J_R at the stable fixed point.

Very Fast Tune Modulation

Very fast tune modulation, with $Q_M \gg Q_I$, corresponds to phase modulation [40, 47]. The resonance Hamiltonian of Equation 16.10 becomes

$$H_{Rn} = \alpha\, I^2 \;-\; V_{Rn} \sum_{k=-\infty}^{\infty} J_k\left(\frac{nq}{Q_M}\right) \cos(n\phi) \tag{16.12}$$

after a canonical transformation has been applied, and after the family of Bessel functions J_k has entered through the use of the identity

$$\cos\left(A + B\cos(C)\right) \equiv \sum_{k=-\infty}^{\infty} J_k(B)\cos(A + kC) \tag{16.13}$$

Each integer value k labels a synchrobetatron sideband at a tune of

$$Q_k = \frac{p}{n} + k\frac{Q_M}{n} \tag{16.14}$$

and labels a sideband island chain that is centred at a sideband action of

$$I_k = k\,\frac{\pi Q_M}{n\alpha} \tag{16.15}$$

Trajectories close to this action have a net motion – small after n modulation periods – that is well described by the sideband Hamiltonian

$$H_{Rnk} = \alpha(I - I_k)^2 \;-\; V_{Rn}\, J_k\left(\frac{nq}{Q_M}\right) \cos(n\phi) \tag{16.16}$$

The Bessel function J_k in this expression is of order 1 if the sideband tune is in the approximate range

$$\frac{p}{n} - q \;<\; Q_k \;<\; \frac{p}{n} + q \tag{16.17}$$

a condition that may be satisfied by multiple values of k, each with a sideband island chain of a similar significant size. The island size shrinks rapidly when $|k|$ increases enough to move Q_k outside this range. Thus, sideband k is predicted to be strong (under fast modulation) if

$$|k| \lesssim \frac{nq}{Q_M} \tag{16.18}$$

For example, $2n$ strong sidebands are predicted when q and Q_M are equal. They may overlap, and cause chaos.

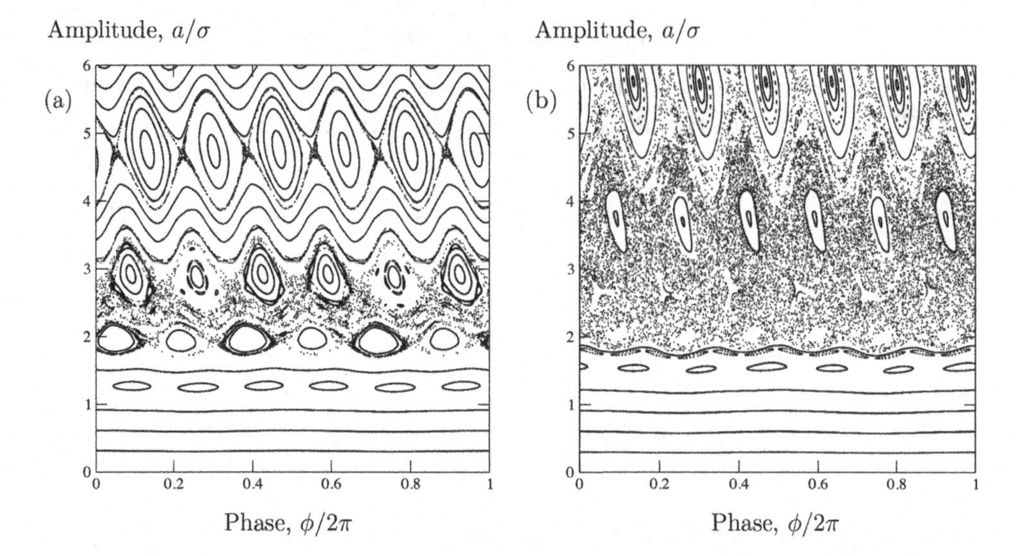

Figure 16.2 Simulated phase space structure due to one round beam–beam kick of strength $\xi = 0.0042$ (a), and $\xi = 0.006$ (b), with the parameters of Equation 16.19. The modest increase in ξ moves the tune modulation sidebands closer together, and dramatically broadens the chaotic sea, allowing rapid transit from amplitudes of 2σ to 6σ and beyond. See also Figure 15.4, which shows the behaviour with no tune modulation.

Three strong sidebands with $k = -1, 0, 1$ are expected in the presence of a single beam–beam interaction with the control parameters

$$\begin{aligned} n &= 6 \\ p &= 2 \\ Q_{00} &= 0.331 \\ q &= 0.001 \\ Q_M &= \frac{1}{194} = 0.00515 \end{aligned} \tag{16.19}$$

Simulation results are shown in Figure 16.2, for two values of ξ. Particles escape quickly across a chaotic sea from about 2σ to 6σ and beyond for a realistic beam–beam parameter as small as $\xi = 0.006$, with only a modest level of tune modulation. Phase space behaviour with no tune modulation is shown for comparison in Figure 15.4.

Very Slow Tune Modulation

Very slow tune modulation, with $Q_M \ll Q_I$, causes amplitude modulation [6, 33, 40, 47]. As the tune Q_0 oscillates, a single resonance island chain moves slowly, first to larger and then to smaller amplitudes. When the modulation is adiabatically slow, individual particles maintain a constant action, or resonance Hamiltonian 'energy'. For example, a particle that initially has an amplitude larger than the resonance island amplitude is squeezed through an unstable fixed point to eventually remain at the same amplitude, if an island chain moves across the particle to a significantly larger amplitude.

A particle that is launched within a resonance island is trapped in that island, and oscillates to larger and smaller amplitudes with the island, if the modulation is adiabatically slow. A bunch of electrons (or a significant fraction of a bunch) that is trapped within an island can be detected by the pattern of synchrotron light that is emitted, a phenomenon first observed in Novosibirsk in 1968 [26]. Similarly, trapped protons generate a persistent signal that is seen on the turn-by-turn output of a beam position monitor, with a frequency that corresponds *exactly* to the resonance tune $Q_0 = p/n$. The strength of this signal increases and decreases with the slow modulation. Resonance island trapping enables nonlinear dynamics experiments in storage rings [7, 8, 29, 33, 48]. It also enables RF gymnastic techniques to manipulate the beam with practical intent [16].

16.3 Dynamical Zones in Tune Modulation Space

Slow tune modulation ($Q_M < Q_I$) is predicted to be adiabatic, and resonance island trapping is possible, if

$$qQ_M < \frac{Q_I^2}{n} \tag{16.20}$$

This condition essentially compares the resonance island speed da_R/dt on the left hand side, with the permissible speed limit on the right hand side [6, 40]. Fast tune modulation ($Q_M > Q_I$) is predicted to generate significantly strong sidebands, according to Equation 16.18, if

$$q > \frac{Q_M}{n} \tag{16.21}$$

Strong sidebands are predicted to overlap and cause chaos if

$$q^{1/4} Q_M^{3/4} < \frac{4}{n\pi^{1/4}} Q_I \tag{16.22}$$

and if Equation 16.21 is also true [9, 40].

These three conditions become universal boundaries that depend only on the resonance order n

$$\left(\frac{q}{Q_I}\right)\left(\frac{Q_M}{Q_I}\right) = \frac{1}{n} \qquad \text{Adiabatic AM or chaos} \tag{16.23}$$

$$\left(\frac{q}{Q_I}\right)\left(\frac{Q_M}{Q_I}\right)^{-1} = \frac{1}{n} \qquad \text{Strong sidebands or not}$$

$$\left(\frac{q}{Q_I}\right)^{1/4}\left(\frac{Q_M}{Q_I}\right)^{3/4} = \frac{4}{(n\pi)^{1/4}} \qquad \text{Overlapped sidebands or not}$$

when the tune modulation amplitude q and tune Q_M are scaled by the island tune Q_I, so that the boundaries are defined in normalised tune modulation space $(q/Q_I, Q_M/Q_I)$. These boundaries apply no matter what the resonance source (beam–beam or magnetic), and no matter what the tune modulation source. The three boundaries, and the four dynamical zones that they separate, are drawn for resonance order $n = 6$ in Figure 16.3. The universality of this perspective emphasises the crucial role played by the island tune, Q_I.

These boundary predictions are readily tested in simulation, for example using a model with three octupoles to provide resonance-free independently variable detuning, and a single decapole to drive $n = 5$ resonance islands [47]. Such tests find

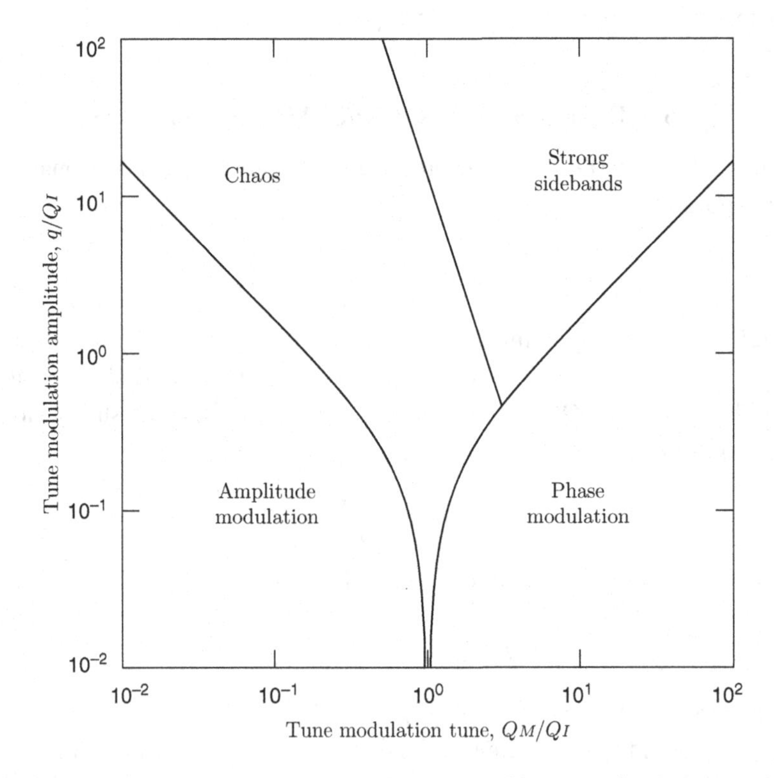

Figure 16.3 Dynamical zones universally predicted in normalised tune modulation space $(q/Q_I, Q_M/Q_I)$ for n = 6, with the boundaries defined in Equation 16.23. The island tune Q_I, a scale factor on both axes, is a parameter of central importance.

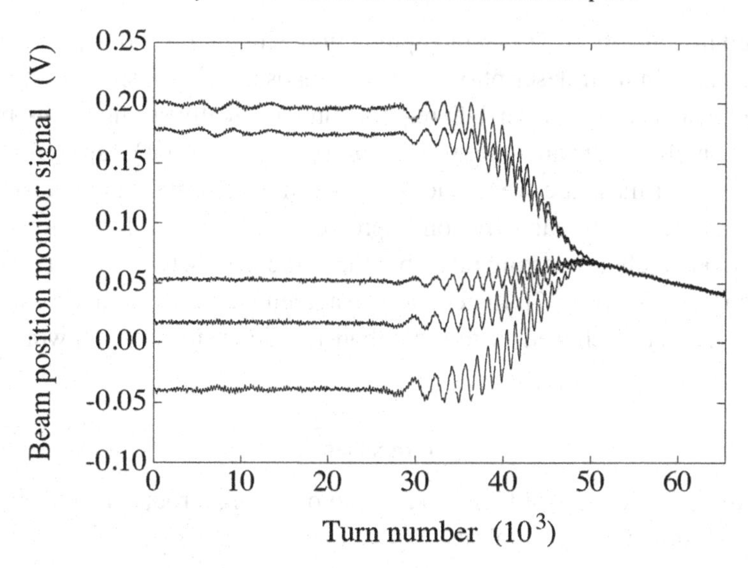

Figure 16.4 Turn-by-turn persistent signal data due to proton beam trapped in an $n = 5$ resonance island in the nonlinear dynamics Tevatron experiment E778 [47, 48]. The resonance is driven by sextupoles. Chirping the operating point from the *amplitude modulation* zone into the *chaos* zone in Figure 16.3 destroys the resonance, and the persistent signal.

that the boundaries are more complex than the simple lines drawn in Figure 16.3. For example, harmonic and sub-harmonic structure at modulation tunes

$$Q_M = \frac{k}{m} Q_I \tag{16.24}$$

(where k and m are integers) show chaotic behaviour even at very small tune modulation amplitudes. Nonetheless, the simple boundaries are good first approximations.

The boundaries can also be tested experimentally. The response of proton beam trapped in an $n = 5$ resonance island to a chirped tune modulation

$$\begin{aligned} q &= q_0 \,(1 \,+\, A(t)) \\ Q_M &= Q_{M0} \,(1 \,+\, A(t)) \end{aligned} \tag{16.25}$$

with a constant ratio q/Q_M is shown in Figure 16.4 [47, 48]. At first, five centre-of-charge locations are observed in a beam position monitor, as the beam hops from island to island, turn-by-turn. The island phases are constant when the operating point is in the amplitude modulation zone, but the island shrinks in size and beam is squeezed out, weakening the persistent signal, as the operating point approaches the boundary between amplitude modulation and chaos. Eventually the signal disappears completely, as the boundary is crossed. This detailed behaviour is successfully reproduced in simulation. Theoretical prediction, simulation and experiment are in good agreement in their description of strong chaos in 1.5 real space dimensions.

Strong resonance overlap in 1-D requires unrealistically strong beam–beam (or magnet) strengths. There are other sources of weak chaos in 1-D, especially Arnol'd diffusion [9], but these are weak enough to be uninteresting. In any case the real world is 3-D. Modulational diffusion is predicted in 2.5 or 3 dimensions [10, 31, 55], with moderately interesting rates, but the agreement between theory and simulation is only qualitative [47]. In general, the agreement between theory, simulation and experiment is much weaker in more than 1.5 dimensions. Much work remains to be done!

Exercises

16.1 Explore the effects of tune modulation on a single round 1-D beam–beam interaction, using a numerical simulation.

a) Reproduce the evolution from regular motion to chaotic motion exhibited in Figures 15.4 and 16.2.
b) Locate the approximate transition to chaos (between Figures 15.4 and 16.2) in the normalised tune modulation space ($Q_M/Q_I, q/Q_I$) plot of Figure 16.3.
c) The modulation period $T_M = 1/Q_M$ must be an integer, so that clear Poincaré surfaces-of-section can be taken every T_M turns. Explore different values of T_M and comment on the behaviour when it has many common factors. For example, compare values 194 and 200.
d) Explore the other boundaries in normalised tune modulation space.

Appendix A

Selected Formulae for Accelerator Design

Also see three standard references: the CERN technical note by Bovet [3], the TRANSPORT user manual [4] and the MADX user manual [17].

A.1 Matrices for Linear Motion through Accelerator Elements

Linear motion from longitudinal location s_1 to s_2 through a generic accelerator element is written

$$\begin{pmatrix} x \\ x' \\ y \\ y' \\ z \\ \delta \end{pmatrix}_2 = M_{21} \begin{pmatrix} x \\ x' \\ y \\ y' \\ z \\ \delta \end{pmatrix}_1 \tag{A.1}$$

where z is the longitudinal displacement of a test particle relative to a reference particle that is usually at the centre of a bunch, and

$$\delta = \frac{\Delta p}{p} \tag{A.2}$$

is the relative momentum offset. The magnet matrices M_{21} given below are in the limit of zero aperture, to make valid the approximation that the field drops to zero instantaneously at the magnet ends.

A.1.1 Field Free Drift

Motion through a field-free region is represented by

$$M_{DRIFT} = \begin{pmatrix} 1 & L & 0 & 0 & 0 & 0 \\ 0 & 1 & 0 & 0 & 0 & 0 \\ 0 & 0 & 1 & L & 0 & 0 \\ 0 & 0 & 0 & 1 & 0 & 0 \\ 0 & 0 & 0 & 0 & 1 & 0 \\ 0 & 0 & 0 & 0 & 0 & 1 \end{pmatrix} \tag{A.3}$$

where L is the length of the drift.

A.1.2 Sector Dipole SBEND

The end faces (and fields) of a pure sector dipole are perpendicular to the design trajectory, on both entrance and exit. If the bending radius is ρ and the bend angle is θ, then

$$M_{SBEND} = \begin{pmatrix} \cos\theta & \rho\sin\theta & 0 & 0 & 0 & \rho(1-\cos\theta) \\ -(\sin\theta)/\rho & \cos\theta & 0 & 0 & 0 & \sin\theta \\ 0 & 0 & 1 & \rho\theta & 0 & 0 \\ 0 & 0 & 0 & 1 & 0 & 0 \\ -\sin\theta & -\rho(1-\cos\theta) & 0 & 0 & 1 & \rho(\theta-\sin\theta) \\ 0 & 0 & 0 & 0 & 0 & 1 \end{pmatrix} \tag{A.4}$$

and $L = \rho\theta$ is the arc length of the design trajectory.

A.1.3 Dipole with Pole-Face Rotations

A dipole with pole-face rotation angles ϕ_1 and ϕ_2 at the entrance and exit is represented by a total matrix

$$M_{DIPOLE} = M_{PFR}(\phi_2)\, M_{SBEND}\, M_{PFR}(\phi_1) \tag{A.5}$$

where

$$M_{PFR}(\phi) = \begin{pmatrix} 1 & 0 & 0 & 0 & 0 & 0 \\ \tan\phi/\rho & 1 & 0 & 0 & 0 & 0 \\ 0 & 0 & 1 & 0 & 0 & 0 \\ 0 & 0 & -\tan\phi/\rho & 1 & 0 & 0 \\ 0 & 0 & 0 & 0 & 1 & 0 \\ 0 & 0 & 0 & 0 & 0 & 1 \end{pmatrix} \tag{A.6}$$

and ρ is the bending radius. A positive angle ϕ rotates the pole-face inwards – in opposite senses at each end – so that a rectangular dipole with parallel ends has

$$\phi_1 = \phi_2 = \theta/2 \tag{A.7}$$

where θ is the bend angle. This expression for M_{PFR} implicitly assumes that the longitudinal profile of the vertical field falls to zero very quickly at the magnet end, equivalent to the assumption that the vertical gap of the dipole is negligibly small [3, 4].

A.1.4 Rectangular Dipole RBEND

A rectangular dipole with parallel end faces and fields has

$$M_{RBEND} = M_{PFR}(\theta/2)\, M_{SBEND}\, M_{PFR}(\theta/2) \tag{A.8}$$

so that

$$M_{RBEND} = \begin{pmatrix} 1 & \rho\theta & 0 & 0 & 0 & \rho(1-\cos\theta) \\ 0 & 1 & 0 & 0 & 0 & \sin\theta \\ 0 & 0 & \cos\theta & \rho\sin\theta & 0 & 0 \\ 0 & 0 & -(\sin\theta)/\rho & \cos\theta & 0 & 0 \\ -\sin\theta & -\rho(1-\cos\theta) & 0 & 0 & 1 & \rho(\theta-\sin\theta) \\ 0 & 0 & 0 & 0 & 0 & 1 \end{pmatrix} \tag{A.9}$$

where ρ is the bending radius and θ is the bend angle.

A.1.5 Sector Combined Function Bend

The matrix representing a sector combined function bending magnet, with ends that are perpendicular to the design trajectory, is

$$M_{SCF} = \begin{pmatrix} C_x & \frac{1}{k_x}S_x & 0 & 0 & 0 & \frac{G}{k_x^2}(1-C_x) \\ -k_xS_x & C_x & 0 & 0 & 0 & \frac{G}{k_x}S_x \\ 0 & 0 & C_y & \frac{1}{k_y}S_y & 0 & 0 \\ 0 & 0 & -k_yS_y & C_y & 0 & 0 \\ -\frac{G}{k_x}S_x & -\frac{G}{k_x^2}(1-C_x) & 0 & 0 & 1 & \frac{G^2}{k_x^3}(k_xL-S_x) \\ 0 & 0 & 0 & 0 & 0 & 1 \end{pmatrix} \tag{A.10}$$

where L is the arc length of the design trajectory, $G = 1/\rho$ is the bending strength, and for brevity

$$\begin{aligned} C_{x,y} &= \cos k_{x,y}L \\ S_{x,y} &= \sin k_{x,y}L \end{aligned} \tag{A.11}$$

The horizontal and vertical focusing strengths

$$k_x^2 = (1 - n)G^2$$
$$k_y^2 = nG^2 \tag{A.12}$$

depend on the dimensionless field index n that describes the relative strength of the quadrupole field component through

$$B_y = B_{y0}\left(1 - \frac{n}{\rho}x \ldots\right) \tag{A.13}$$

where B_y is the vertical field at a horizontal displacement x on the mid-plane. A pure dipole field has $n = 0$, $k_x = G$, and $k_y = 0$.

A.1.6 Quadrupole

The matrix of a horizontally focusing (vertically defocusing) quadrupole is

$$M_{QUAD} = \begin{pmatrix} \cos kL & \frac{1}{k}\sin kL & 0 & 0 & 0 & 0 \\ -k\sin kL & \cos kL & 0 & 0 & 0 & 0 \\ 0 & 0 & \cosh kL & \frac{1}{k}\sinh kL & 0 & 0 \\ 0 & 0 & k\sinh kL & \cosh kL & 0 & 0 \\ 0 & 0 & 0 & 0 & 1 & 0 \\ 0 & 0 & 0 & 0 & 0 & 1 \end{pmatrix} \tag{A.14}$$

where both

$$k = \sqrt{K} \tag{A.15}$$

and the local quadrupole strength

$$K = \mathrm{sgn}(q)\,\frac{B'}{(B\rho)} \tag{A.16}$$

are positive-definite, but the sign of $B' = dB_y/dx$ depends on the sign of q, the charge of the particle.

The horizontal and vertical 2×2 matrix sub-blocks are exchanged when the quadrupole is horizontally defocusing, in which case

$$k = \sqrt{-K} \tag{A.17}$$

is positive and K is negative.

A.1.7 Rolled Magnet

The matrix for a magnet that has been rolled clockwise by an angle α about the longitudinal axis is

$$M_{ROLLED} = R(-\alpha)\, M\, R(\alpha) \tag{A.18}$$

where M is the matrix of the unrolled magnet, R is the co-ordinate transformation

$$R = \begin{pmatrix} C & 0 & S & 0 & 0 & 0 \\ 0 & C & 0 & S & 0 & 0 \\ -S & 0 & C & 0 & 0 & 0 \\ 0 & -S & 0 & C & 0 & 0 \\ 0 & 0 & 0 & 0 & 1 & 0 \\ 0 & 0 & 0 & 0 & 0 & 1 \end{pmatrix} \tag{A.19}$$

and

$$C = \cos\alpha \tag{A.20}$$
$$S = \sin\alpha$$

For example

$$M_{SKEW\ QUAD} = R(-\pi/4)\, M_{QUAD}\, R(\pi/4) \tag{A.21}$$

represents a thick skew quadrupole that has been rolled by 45 degrees. In general M_{ROLLED} is not block-diagonal, and so it couples horizontal and vertical motion.

A.1.8 Solenoid

A solenoid with a purely longitudinal field B_s is represented by a matrix

$$M_{SOLENOID} = \begin{pmatrix} C^2 & \frac{1}{K}SC & SC & \frac{1}{K}S^2 & 0 & 0 \\ -K\,SC & C^2 & -K\,S^2 & SC & 0 & 0 \\ -SC & -\frac{1}{K}S^2 & C^2 & \frac{1}{K}SC & 0 & 0 \\ K\,S^2 & -SC & -K\,SC & C^2 & 0 & 0 \\ 0 & 0 & 0 & 0 & 1 & 0 \\ 0 & 0 & 0 & 0 & 0 & 1 \end{pmatrix} \tag{A.22}$$

where

$$K = \mathrm{sgn(q)}\,\frac{B_s}{2(B\rho)} \tag{A.23}$$

is the local geometric strength of the solenoid, and

$$C = \cos KL \tag{A.24}$$
$$S = \sin KL$$

where L is the solenoid length. This matrix is only block-diagonal when $S = 0$, so in general solenoids couple horizontal and vertical motion.

A.1.9 Acceleration in the Relativistic Limit

In the fully relativistic limit $\gamma \to \infty$ the matrix representing an accelerating section of length L is

$$M_{RF} = \begin{pmatrix} 1 & L\frac{\ln(1+\Delta)}{\Delta} & 0 & 0 & 0 & 0 \\ 0 & \frac{1}{1+\Delta} & 0 & 0 & 0 & 0 \\ 0 & 0 & 1 & L\frac{\ln(1+\Delta)}{\Delta} & 0 & 0 \\ 0 & 0 & 0 & \frac{1}{1+\Delta} & 0 & 0 \\ 0 & 0 & 0 & 0 & 1 & 0 \\ 0 & 0 & 0 & 0 & \frac{2\pi}{\lambda}\frac{t\Delta}{1+\Delta} & \frac{1}{1+\Delta} \end{pmatrix} \tag{A.25}$$

where

$$\Delta = \frac{\Delta E \cos\phi}{E_0} \tag{A.26}$$

is the relative energy gain of a reference test particle with an initial energy of E_0 that passes through an accelerating section with a maximum possible energy gain of ΔE, with an RF phase lag ϕ. The wavelength of the RF system is λ, and

$$t = \tan\phi \tag{A.27}$$

where a positive phase lag ϕ means that a test particle with a positive z (at the head of the bunch) is accelerated more than the reference particle.

A.2 Propagation of Twiss Functions

A.2.1 Difference Equations

Twiss functions propagate from s_1 to s_2 through

$$\begin{pmatrix} \beta_2 \\ \alpha_2 \\ \gamma_2 \end{pmatrix} = \begin{pmatrix} m_{11}^2 & -2m_{11}m_{12} & m_{12}^2 \\ -m_{21}m_{11} & 1+2m_{12}m_{21} & -m_{12}m_{22} \\ m_{21}^2 & -2m_{22}m_{21} & m_{22}^2 \end{pmatrix} \begin{pmatrix} \beta_1 \\ \alpha_1 \\ \gamma_1 \end{pmatrix} \tag{A.28}$$

and the phase advance is found by inverting the equation

$$\tan(\phi_2 - \phi_1) = \frac{m_{12}}{m_{11}\beta_1 - m_{12}\alpha_1} \tag{A.29}$$

where

$$M_{21} = \begin{pmatrix} m_{11} & m_{12} \\ m_{21} & m_{22} \end{pmatrix} \tag{A.30}$$

represents transverse motion from s_1 to s_2 in the plane of interest.

A.2.2 Differential Equation

The Twiss β-function propagates via the differential equation

$$\frac{d^2 b}{ds^2} + K b - b^{-3} = 0 \tag{A.31}$$

where $b = \sqrt{\beta}$ and $K(s)$ is the local quadrupole strength, with a different sign in horizontal and vertical planes.

A.2.3 Propagation through a Drift

If s_1 and s_2 are separated by a drift of length L then

$$\begin{aligned} \beta_2 &= \beta_1 - 2\alpha_1 L + \gamma_1 L^2 \\ \alpha_2 &= \alpha_1 - \gamma_1 L \\ \gamma_2 &= \gamma_1 \\ \phi_2 &= \phi_1 + \arctan\left(\frac{L}{\beta_1 - \alpha_1 L}\right) \\ \phi_2 &= \phi_1 + \arctan\left(\frac{L}{\sqrt{\beta_1 \beta_2}}\right) \end{aligned} \tag{A.32}$$

This is solved by

$$\begin{aligned} \beta(s) &= \beta^* + \frac{(s - s_0)^2}{\beta^*} \\ \alpha(s) &= -\frac{(s - s_0)}{\beta^*} \\ \gamma(s) &= \frac{1}{\beta^*} \end{aligned} \tag{A.33}$$

for some pair of values, s_0 and β^*.

References

[1] V. Balakin, A. Novokhatsky, and V. Smirnov, VLEPP: transverse beam dynamics. In: *PAC83*, 1983.

[2] A. Ben-Israel, A note on an iterative method for generalized inversion of matrices. *Math. Comput.*, **20** (1966), 439.

[3] C. Bovet, R. Gouiran, I. Gumowski, and K.H. Reich, *A Selection of Formulae and Data Useful for the Design of A.G. Synchrotrons*. Tech. rept. CERN/MPS-SI/Int. DL/70/4. CERN, 1970.

[4] K.L. Brown, D.C. Carey, Ch. Iselin, and F. Rothacker, *TRANSPORT: A Computer Program for Designing Charged Particle Beam Transport systems*. Tech. rept. SLAC-91, NAL-91, CERN-80-04. SLAC, Fermilab, CERN, 1983.

[5] A. Chao, *Physics of Collective Instabilities in High Energy Accelerators*. Wiley, 1993.

[6] A. Chao and M. Month, Particle trapping during passage through a high-order nonlinear resonance. *Nuclear Instruments and Methods*, **121** (1974), 129–138.

[7] A. Chao, D. Johnson, S. Peggs, J. Peterson, C. Saltmarsh, L. Schachinger, R. Meller, R. Siemann, R. Talman, P. Morton, D. Edwards, D. Finley, R. Gerig, N. Gelfand, M. Harrison, R. Johnson, N. Merminga, and M. Syphers, Experimental investigation of nonlinear dynamics in the Fermilab Tevatron. *Physical Review Letters*, **61**:24 (1988), 2752–2756.

[8] T. Chen, A. Gerasimov, B. Cole, D. Finley, G. Goderre, M. Harrison, R. Johnson, I. Kourbanis, C. Manz, N. Merminga, L. Michelotti, S. Peggs, F. Pilat, S. Pruss, C. Saltmarsh, S. Saritepe, T. Satogata, R. Talman, C.G. Trahern, and G. Tsironis, Measurements of a Hamiltonian system and their description by a diffusive model. *Physical Review Letters*, **68**:1 (1992), 33–37.

[9] B. Chirikov, A universal instability of many-dimensional oscillator systems. *Physics Reports*, **52**:5 (1979), 263–379.

[10] B. Chirikov, M. Lieberman, D. Shepelyansky, and F. Vivaldi, A theory of modulational diffusion. *Physica D*, **14**:3 (1985), 289–304.

[11] M. Conte and W.M. MacKay, *An Introduction to the Physics of Particle Accelerators*. World Scientific, 1991.

[12] H.S. Dumas, *The KAM Story*. World Scientific, 2014.

[13] D. Edwards, *Oscillation Damping Factors for Off-Momentum Orbits in Electron Storage Rings*. Tech. rept. FNAL TM-566 1501. Fermilab, 1975.

[14] D. Edwards and L. Teng, Parameterization of linear coupled motion in periodic systems. *IEEE Transactions of Nuclear Science*, **20**:3 (1973), 885–888.

[15] W. Fischer, Private communication, 2014.

[16] W. Gabella, J. Rosenzweig, R. Kick, and S. Peggs, RF voltage modulation at discrete frequencies with applications to crystal channeling extraction. *Particle Accelerators*, **42** (1993), 235–257.

[17] H. Grote, F. Schmidt, L. Deniau, and G. Roy, *MAD - Methodical Accelerator Design*, 2015.

[18] M. Harrison, S. Peggs, and T. Roser, The RHIC accelerator. *Annual Review of Nuclear and Particle Science*, **52** (2002), 425–469.

[19] M. Hénon, Numerical study of quadratic area-preserving mappings. *Quarterly of Applied Math*, **27**:3 (1969), 291.

[20] D.R. Hofstadter, *Metamagical Themas: Questing for the Essence of Mind and Pattern*. Basic Books, 1985.

[21] J.D. Jackson, *Classical Electrodynamics*. John Wiley and Sons, 1998.

[22] J. Johnstone, *A Simplified Analysis of Resonant Extraction at the Main Injector; A Numerical Simulation of Resonant Extraction*. Tech. rept. MI-0091; MI-0095. Fermilab, 1993.

[23] I.M. Kapchinsky and V.A. Teplyakov, Linear ion accelerator with spatially homogeneous strong focusing. *Pribory i Tekhnika Eksperimenta*, **2** (1970), 19–22.

[24] E. Keil, *Beam-Beam Dynamics*. Tech. rept. CERN-SL-94-78-AP. CERN, 1994.

[25] Y. Kobayashi, Theory of the resonant beam ejection from synchrotrons. *Nuclear Instruments and Methods*, **83** (1970), 77–87.

[26] G. Kulipanov, S. Mishnev, S. Popov and G. Tumaikin, Influence of nonlinearities in driven betatron oscillations. *Novosibirsk Preprint INP*, **68**:251 (1968).

[27] J. Le Duff, Dynamics and acceleration in linear structures. In: *CAS-CERN Accelerator School, 1992 (CERN-1994-001)*, 1994.

[28] V. Lebedev, N. Solyak, J.-F. Ostiguy, A. Alexandrov, and A. Shishlo, Intrabeam stripping in H^- linacs. In: *LINAC10*, 2010.

[29] S.Y. Lee, M. Ball, B. Brabson, D. Caussyn, J. Collins, S. Curtis, V. Derenchuck, D. DuPlantis, G. East, M. Ellison, T. Ellison, D. Friesel, B. Hamilton, W.P. Jones, W. Lamble, D. Li, M. Minty, P. Schwandt, T. Sloan, G. Xu, A. Chao, S. Tepikian, and K.Y. Ng, Experimental determination of a nonlinear Hamiltonian in a synchrotron. *Physical Review Letters*, **67**:27 (1991), 3768–3771.

[30] C. Leemann, D. Douglas, and G. Krafft, The continuous electron beam accelerator facility: CEBAF at the Jefferson Laboratory. *Annual Review of Nuclear and Particle Science*, **51** (2001), 413–450.

[31] A. Lichtenberg and M. Lieberman, *Regular and Stochastic Motion*. Springer-Verlag, 1983.

[32] L. Merminga, D. Douglas, and G. Krafft, High-Current energy-recovering electron linacs. *Annual Review of Nuclear and Particle Science*, **53** (2003), 387–429.

[33] N. Merminga, *A Study of Nonlinear Dynamics in the Fermilab Tevatron*. Ph.D. thesis, University of Michigan, 1989.

[34] K.-H. Mess, P. Schmüser, and S. Wolff, *Superconducting Accelerator Magnets*. World Scientific, 1996.

[35] J.B. Murphy, *Synchrotron Light Source Data Book*. http://scitation.aip.org/content/aip/proceeding/aipcp/10.1063/1.41969: BNL Report 42333, 1996.

[36] H. Padamsee, J. Knobloch, and T. Hays, *RF Superconductivity for Accelerators*. Wiley-VCH, 2008.

[37] V. Pan and J. Reif, In: *Proc. of 17th Ann. ACM Symp. on Theory of Computing*, 1985.

[38] S. Peggs, *Some Aspects of Machine Physics in the Cornell Electron Storage Ring*. Ph.D. thesis, Cornell University, 1981.

[39] S. Peggs, Coupling and decoupling in storage rings. *IEEE Transactions of Nuclear Science*, **30**:4 (1983), 2460–2462.
[40] S. Peggs, *Hamiltonian Theory of the E778 Nonlinear Dynamics Experiment*. Tech. rept. SSC-175; CERN 88-04. SSC, CERN, 1988.
[41] S. Peggs, Feedback between accelerator physicists and magnet builders. In: *Proc. of the LHC Single Particle Dynamics Workshop, Montreux; also RHIC/AP/80*, 1995.
[42] S. Peggs and J. Wei, *Longitudinal Phase Space Parameters*. Tech. rept. RHIC/AP/106. BNL, 1996.
[43] S. Peggs and R.M. Talman, Nonlinear problems in accelerator physics. *Annual Review of Nuclear and Particle Science*, **36** (1986), 287–325.
[44] H. Poincaré, *Les Methods Nouvelle de la Mechanique Celestes*. Gautier-Vilars, 1892.
[45] E. Pozdeyev, Regenerative multipass beam breakup in two dimensions. *Physical Review Accelerators and Beams*, **8**:054401 (2005), 1–17.
[46] M. Sands, *The Physics of Electron Storage Rings*. Tech. rept. SLAC-121. SLAC, 1970.
[47] T. Satogata, *Nonlinear Resonance Islands and Modulational Effects in a Proton Synchrotron*. Ph.D. thesis, Northwestern University, 1993.
[48] T. Satogata, T. Chen, B. Cole, D. Finley, A. Gerasimov, G. Goderre, M. Harrison, R. Johnson, I. Kourbanis, C. Manz, N. Merminga, L. Michelotti, S. Peggs, F. Pilat, S. Pruss, C. Saltmarsh, S. Saritepe, R. Talman, C.G. Trahern, and G. Tsironis, Driven response of a trapped particle beam. *Physical Review Letters*, **68**:12 (1992), 1838–1841.
[49] F. Schmidt and F. Willeke, Nonlinear beam dynamics close to resonances excited by sextupole fields. In: *EPAC88*, 1988.
[50] SLAC *Linac Coherent Light Source II (LCLS-II) Conceptual Design Report*. Tech. rept. SLAC-R-978. SLAC, 2011.
[51] P. Stewart, Jacobellis v. Ohio. *U.S. Supreme Court*, 1964.
[52] L.G. Taff, *Celestial Mechanics*. Wiley, 1985.
[53] C. Tennant, *'Energy Recovery Linacs', in Challenges and Goals for Accelerators in the XXI Century*. World Scientific, 2016.
[54] D. Trbojevic and M. Harrison, Design and multiparticle simulation of the half-integer slow extraction system for the Main Injector. In: *PAC91*, 1991.
[55] F. Vivaldi, Weak instabilities in many-dimensional Hamiltonian systems. *Reviews of Modern Physics*, **56**:4 (1984), 737–755.
[56] M. Vretenar, The radio-frequency quadrupole. *CERN-2013-001*, 207–223, 2013.
[57] T. Wangler, *RF Linear Accelerators*. Wiley-VCH, 2008.
[58] J. Wei, *Longitudinal Dynamics of the Non-Adiabatic Regime of Alternating Gradient Synchrotrons*. Ph.D. thesis, Stony Brook University, 1990.
[59] Wikipedia. *List of Accelerators in Particle Physics*.
[60] Wikipedia. *List of Synchrotron Radiation Facilities*.

Index

Printed in the United States
by Baker & Taylor Publisher Services